QUR'AN
—AND—
WOMAN

AMINA WADUD-MUHSIN

KUALA LUMPUR
PENERBIT FAJAR BAKTI SDN. BHD.

Penerbit Fajar Bakti Sdn. Bhd.
19-25, Jalan Kuchai Lama,
58200 Kuala Lumpur

© Penerbit Fajar Bakti Sdn. Bhd. 1992
First published 1992
Third impression 1994

ISBN 967 65 1976 6

Printed in Malaysia by
Percetakan Direct Art Sdn. Bhd., Kuala Lumpur

Dedication

This book is dedicated to all who struggle to hear the voice of their faith expressed by the Qur'an—especially the Sisters in Islam.

Preface

*T*HIS is an analysis of the concept of woman drawn directly from the Qur'an. The Qur'an was a political, social, spiritual, and intellectual catalyst for change in the lives of the immediate community of people in the Arabian peninsula. Its influence expanded into a large geographical area in an unusually short period of time. The expansion of the ideology espoused by the Qur'an remains in effect in all areas to which it spread, with the exception of Spain. Therefore, the Qur'an represents a 'world-altering' force which must be recognized and understood.

In the first century after the spread of Islam and the widespread acceptance of the Qur'an, it began to play its most significant role in the intellectual arena. As the desire to understand the text grew, a number of interpretative methodologies developed. These methods reflected the objectives of a number of disciplines, which in turn, developed into distinct 'sciences' or categories of study within Islamic scholarship. *Shari'ah* (Islamic Law), Grammar, Literature and Politics are some of the more important disciplines whose developments were based on the Qur'an.

Each of these disciplines generated a great deal of literature. As this literature developed, it began to play a role so central in Islamic scholarship that it overshadowed the text upon which it was originally based. Consequently, Islamic studies began to focus more heavily on the understanding of this literature and less on the understanding of the Qur'an itself. The result was a disconnection from the original text and its intent.

This study involves an explicit attempt to overcome the disconnection from that original source—the motivating force in the development of such copious literature. Yet, there is an even more significant rationale behind this attempt to overcome the disconnection with the original. As the various disciplines developed, they satisfied particular needs or answered particular questions that had arisen. Thus the forces that motivated each discipline also defined the parameters in which that discipline was to focus. In the present day, it has become apparent that certain questions did not arise. Consequently, the answers to these were not sought out. Obviously, the answers to these were not found.

The question of the *concept* of woman in the Qur'an did not arise—perhaps because the concept of gendered man did not arise. The critical questioning of the functions and responsibilities of each gender has only recently been asked: inspired, for the most part, by the sad condition of women in Islamic societies at the time of independence from colonialist forces. Once the question arose, a sound method of answering that question needed to be developed within the field of Islamic scholarship.

One significant aspect of that method would be to overcome the disconnection with the original source of guidance for all humankind, the Qur'an. Yet, this, of all aspects has been overlooked. Instead, the tendency has been to confuse the works of Islamic scholars (past and present) with the Qur'an—which supposedly motivated those works. In recent years a new confusion has arisen which pits non-Islamic sources with the original sources of Islam. Thus, scholars have attempted to draw principles from non-Islamic sources and apply them to the Qur'an, rather than to extract Qur'anic principles and apply them to particular problems.

Finally, this particular issue of woman in the Qur'an has two significant implications. First, it will demonstrate in specific terms my proposal that in order to maintain its relevance, the Qur'an must be continually re-interpreted. Second, the progress of civilization has been reflected in the extent of the woman's participation in society and the recognition of the significance of her resources.[1] In the context of Islam and the Muslims, the understanding of the Qur'anic concept of woman established over 1,400 years ago indicates the advanced level of that civilization. If it had been fully implemented in the practical sense, then Islam would have been a global motivating force for women's empowerment.

That I have chosen to consider the significance of a fourteen-hundred-year-old text to that issue is unique only with regard to some feminists' perspectives which require a complete break from the past. For these women, the maximization of progress is directly related to freedom from binding 'traditions'. The importance of the Qur'anic text is its transcendence of time and its expression of eternal values. As such, the context of Muslim communities has not yet risen to the level of the text. It was not the text which restricted woman, but the interpretations of that text which have come to be held in greater importance than the text itself. In other religions, feminists have had to insert woman into the discourse in order to attain legitimacy. The Muslim woman has only to read the text—unconstrained by exclusive and restrictive interpretations—to gain an undeniable liberation.

Therefore, the significance of this analysis of the Qur'anic concept of woman is measured within the perspective of the text itself, both as a force in history, politics, language, culture, intellect and spirit, and as a divine text claiming to present guidance for all humankind. Through reviewing the Qur'an itself—with its principles of social justice and human equality, and its objective of guidance—I hope to shed some new light on the role of woman. That this issue has been discussed from other perspectives only reminds us of its significance in the modern context.

January 1992　　　　　　　　　　　　　　　AMINA WADUD-MUHSIN
Kuala Lumpur

---

[1]Yvonne Haddad, 'Islam, Women and Revolution in Twentieth-Century Arab Thought', Yvonne Haddad and Ellison B. Findley, (eds.), *Women, Religion and Social Change* (New York: State University of New York Press, 1985) p. 275.

Acknowledgements

*T*HIS work represents two stages of development. The initial research was taken as part of my graduate studies at the University of Michigan between 1986 and 1989. Although that original project did not meet with much discouragement, neither was it met with enthusiastic encouragement. The one notable exception, whom I wish to acknowledge here, was my 'favourite' professor, Dr Alton Becker (Pete).

In another volume I have acknowledged that access to Pete's private office on campus provided me with a solitary spot in which to do my research. However, I am indebted to him for two general insights which he uttered sometime during my studies with him (1982-9).

First, he mentioned that some 'professors' were unable to meet students and former students as colleagues (what a senior professor in my first university-teaching post described as 'horizontal communication'), which thwarted productive dialogue. Thus began my own development for transcending hierarchy. I have carried the memory of this statement so that I might learn from, and perhaps teach people from, a variety of stations in life.

His second comment was a description of one professor as a man who worked on 'dead texts'. Thus began the central idea in this present work: despite fourteen centuries of existence, the Qur'an must be kept alive. Otherwise, it will suffer the fate of other 'dead texts' and defeat its stated purpose: to guide humankind—unconstrained by time or place.

When I came to Malaysia in 1989, one of the first significant relationships I had the pleasure of developing was with Dr Chandra Muzaffar. I am heavily indebted to him for his suggestions on this present work. Although busy with his own job, research, and public services, he has given considerable time to my first draft and subsequent typescripts. In addition to the benefit of his editorial suggestions, I am grateful for his moral encouragement. I hope the significance he saw in this work might actually benefit Muslims in their future concerns.

Moreover, in the global context of human discourse, I hope one day to attain his calibre of scholarship and sincere commitment to constructive self-criticism.

Finally, I wish to acknowledge the sacrifices of my children in the six years of my involvement with this one project. They will inherit this earth after me and still be required to make a home of it. Apart from my two sons, Muhammad and Khalil-Allah, I wish especially to acknowledge my daughters, Hasnaa, Sahar, and Alaa, as sisters in Islam.

Contents

Introduction
How Perceptions of Woman Influence Interpretation of the Qur'an

*M*Y objective in undertaking this research was to make a 'reading' of the Qur'an that would be meaningful to women living in the modern era. By 'reading' I mean the process of reviewing the words and their context in order to derive an under-standing of the text. Every 'reading' reflects, in part, the intentions of the text, as well as the 'prior text'[1] of the one who makes the 'reading'. Although each 'reading' is unique, the understanding of various readers of a single text will converge on many points.

In this Introduction I will give the background to this work. In particular, I will look at how the perception of woman influences the interpretations of the Qur'an's position on women. I will give an overview of my own perspective of woman and of the methods of interpretation I used in analysing the Qur'an which have led to some new conclusions.

No method of Qur'anic exegesis is fully objective. Each exegete makes some subjective choices. Some details of their interpretations reflect their subjective choices and not necessarily the intent of the text. Yet, often, no distinction is made between text and interpreta-tion. I put interpretations of woman in the Qur'an into three cat-egories: 'traditional', reactive, and holistic.

The first category of Qur'anic interpretation I call 'traditional'. Traditional *tafsir* (exegetical works) give interpretations of the en-tire Qur'an, whether from the modern or classical periods, with cer-tain objectives in mind. Those objectives could be legal, esoteric, grammatical, rhetorical, or historical. Although these objectives

1

may lead to differences in the *tafasir*, one similarity in these works is their atomistic methodology. They begin with the first verse of the first chapter and proceed to the second verse of the first chapter—one verse at a time—until the end of the Book. Little or no effort is made to recognize themes and to discuss the relationship of the Qur'an to itself, thematically. A brief mention of one verse's relation to another verse may be rendered but these are haphazard with no underlying hermeneutical principle applied. A methodology for linking similar Qur'anic ideas, syntactical structures, principles, or themes together is almost non-existent.[2]

However, what concerns me most about 'traditional' *tafasir* is that they were exclusively written by males. This means that men and men's experiences were included and women and women's experiences were either excluded or interpreted through the male vision, perspective, desire, or needs of woman.[3] In the final analysis, the creation of the basic paradigms through which we examine and discuss the Qur'an and Qur'anic interpretation were generated without the participation and firsthand representation of women. Their voicelessness during critical periods of development in Qur'anic interpretation has not gone unnoticed, but it has been mistakenly equated with voicelessness in the text itself.

The second category of Qur'anic interpretation concerned with the issue of woman consists primarily of modern scholars' reactions to severe handicaps for woman as an individual and as a member of society which have been attributed to the text. In this category are many women and/or persons opposed to the Qur'anic message (or more precisely, to Islam) altogether. They use the poor status of women in Muslim societies as justification for their 'reactions'. These reactions have also failed to draw a distinction between the interpretation and the text.[4]

The objectives sought and methods used, often come from feminist ideals and rationales. Although they are often concerned with valid issues, the absence of a comprehensive analysis of the Qur'an sometimes causes them to vindicate the position of women on grounds entirely incongruous with the Qur'anic position on woman. This shortcoming must be overcome in order to make use of a most effective tool for the liberation of Muslim women: demonstrating the link between that liberation and this primary source of Islamic ideology and theology.

The interpretations which reconsider the whole method of Qur'anic exegesis with regard to various modern social, moral, economic, and political concerns—including the issue of woman—represent the final category. It is in this category that I place this work. This category is relatively new, and there has been no substantial consideration of the particular issue of woman in the light of the entire Qur'an and its major principles.

I propose to make a 'reading' of the Qur'an from within the female experience and without the stereotypes which have been the framework for many of the male interpretations. In the final analysis, this reading will confront some of the conclusions drawn on this subject. Because I am analysing the text and not the interpretations of that text, my treatment of this issue differs from many of the existing works on this topic.

Background

Methodology: A Hermeneutical Model

A hermeneutical model is concerned with three aspects of the text, in order to support its conclusions: **1.** the context in which the text was written (in the case of the Qur'an, in which it was revealed); **2.** the grammatical composition of the text (how it says what it says); and **3.** the whole text, its *Weltanschauung* or world-view. Often, differences of opinion can be traced to variations in emphasis between these three aspects.

I argue against some conventional interpretations, especially about certain words used in the Qur'an to discuss and fulfil universal guidance. I render some discussions heretofore considered as gendered, into neutral terms. Other discussions, heretofore considered as universal, I render specific on the basis of their limitations and on the expression in terms specific to seventh-century Arabia. Some historical information with regard to occasions of revelation and the general period of revelation was considered here.

Thus, I attempt to use the method of Qur'anic interpretation proposed by Fazlur Rahman. He suggests that all Qur'anic passages, revealed as they were in a specific time in history and within certain general and particular circumstances, were given expression relative to those circumstances. However, the message is not limited to that time or those circumstances historically. A reader must understand the implications of the Qur'anic expressions during the time in which they were expressed in order to determine their proper meaning. That meaning gives the intention of the rulings or principles in the particular verse.

Believers from another circumstance must make practical applications in accordance with how that original intention is reflected or manifested in the new environments. In modern times this is what is meant by the 'spirit' of the Qur'an. To get at that 'spirit', however, there must be some comprehensible and organized hermeneutical model.[5]

The initial question behind my research was, why does the Qur'an specify males and females on some occasions (like 'Believing males and Believing females' [masculine plural followed by feminine plural forms]), while on other occasions it uses a more generic ('Oh you who believe ...' [masculine plural]) form? From my perspective on the Qur'an, every usage of the masculine plural form is intended to include males and females, *equally*, unless it includes specific indication for its exclusive application to males.

The plural in Arabic is used to denote three or more rational beings. Thus the following Arabic sentences:

A. *Al-tullab fi al-ghurfah* (masculine plural form) means
　　1. three or more students in the room—including at least one male
　　2. three or more *exclusively* male students in the room.
B. *Al-talibat fi al-ghurfah* (feminine plural form) means
　　1. three or more female students in the room.

As there is no form exclusively for males, the only way to determine if the masculine plural form (*al-tullab fi al-ghurfah* (**A**)) is exclusively for male (2) would be through some specific indication in the text. Thus:

C. *Al-tullab wa al-talibat fi al-ghurfah* indicates that the use of the

4

masculine plural (*al-tullab*) refers *exclusively* to males since the in-clusion of the female plural form distinguishes the female students present.[6]

All the verses which contained any reference to women, separately or together with men, were analysed with the traditional method of *tafsir al Qur'an bi al Qur'an* (interpretation of the Qur'an based on the Qur'an itself). However, I elaborated these particular terms of this method: each verse was analysed: **1.** in its context; **2.** in the context of discussions on similar topics in the Qur'an; **3.** in the light of similar language and syntactical structures used elsewhere in the Qur'an; **4.** in the light of overriding Qur'anic principles; and **5.** within the context of the Qur'anic *Weltanschauung*, or world-view.

Language and Prior Text

One unique element for reading and understanding any text is the prior text of the individual reader: the language and cultural context in which the text is read. It is inescapable and represents, on the one hand, the rich varieties that naturally occur between readers, and, on the other hand, the uniqueness of each.

Prior text adds considerably to the perspective and conclusions of the interpretation. It exposes the individuality of the exegete. This is neither good nor bad in and of itself. However, when one individual reader with a particular world-view and specific prior text asserts that his or her reading is the only possible or permissible one, it prevents readers in different contexts to come to terms with their own relationship to the text.

To avoid the potential of relativism, there is continuity and per-manence in the Qur'anic text itself as exemplified even through va-rious readings by their points of convergence. However, in order for the Qur'an to achieve its objective to act as a catalyst affecting behaviour in society, each social context must understand the fun-damental and unchangeable principles of that text, and then imple-ment them in their own unique reflection. It is not the text or its principles that change, but the capacity and particularity of the un-derstanding and reflection of the principles of the text within a community of people.

Thus, each individual reader interacts with the text. However, the assertion that there is only one interpretation of the Qur'an limits the extent of the text. The Qur'an must be flexible enough to accommodate innumerable cultural situations because of its claims to be universally beneficial to those who believe.[7] Therefore, to force it to have a single cultural perspective—even the cultural perspective of the original community of the Prophet—severely limits its application and contradicts the stated universal purpose of the Book itself.

THE PRIOR TEXT OF GENDER-SPECIFIC LANGUAGES

The significance of masculine and feminine forms, whether used distinctively or to make generic indications, was an important part of my analysis. Perspectives on gender, particularly on the understanding of what constitutes feminine or masculine behaviour, and the roles of men and women in society, are based on one's cultural context. Gender-specific languages, such as Arabic, create a particular prior text for the speakers of that language. Everything is classified male or female. English, Malay, and other languages do not share this prior text with Arabic. This results in a distinction between the various readings of the Qur'an. This distinction becomes apparent in the interpretation of the text and the conclusions drawn from the function of the text with regard to gender.

With regard to Arabic, the language of the Qur'an, I approach the text from the outside. This frees me to make observations which are not imprisoned in the context of a gender-distinct language.

There exists a very strong, but one-sided and thus untrustworthy, idea that in order better to understand a foreign culture, one must enter into it, forgetting one's own, and view the world through the eyes of this foreign culture. This idea, as I have said, is one-sided. Of course, a certain entry as a living being into a foreign culture, the possibility of seeing the world through its eyes, is a necessary part of the process of understanding it; but if this were the only aspect of this understanding, *it would merely be duplication and would not entail anything new or enriching.* Creative understanding does not renounce itself, its own place in time, its own culture; and it forgets nothing. In order to understand, it is immensely important for the person who understands to be located outside the object of his or her creative understanding—in time, in space, and in culture.[8] [emphasis mine]

A new look at Qur'anic language with regard to gender is especially necessary in the light of the absence of an Arabic neuter.

Although each word in Arabic is designated as masculine or feminine, it does not follow that each use of masculine or feminine persons is necessarily restricted to the mentioned gender—from the perspective of universal Qur'anic guidance.[9] A divine text must overcome the natural restrictions of the language of human communication. Those who argue that the Qur'an cannot be translated believe that there is some necessary correlation between Arabic and the message itself. I will demonstrate that gender distinction, an inherent flaw, necessary for human communication in the Arabic, is overcome by the text in order to fulfil its intention of universal guidance.

Perspectives on Women

'Most men have at one time or another heard, or perhaps even believed, that women are "inferior" and "unequal" to men.'[10] I worked against the backdrop of common prejudices and attitudes among Muslims towards women which have not only affected the position of women in Muslim societies but also affected the interpretation of the position of women in the Qur'an. One such belief is that there are *essential* distinctions between men and women reflected in creation, capacity and function in society, accessibility to guidance (particularly to Qur'anic guidance), and in the rewards due to them in the Hereafter.

Although there are distinctions between women and men, I argue that they are not of their essential natures. More importantly, I argue against the *values* that have been attributed to these distinctions. Such attributed values describe women as weak, inferior, inherently evil, intellectually incapable, and spiritually lacking. These evaluations have been used to claim that women are unsuitable for performing certain tasks, or for functioning in some ways in society.

The woman has been restricted to functions related to her biology. The man, on the other hand, is evaluated as superior to and more significant than woman, an inherent leader and caretaker, with extensive capacity to perform tasks that the woman cannot. Consequently, men are *more* human, enjoying completely the choice of movement, employment, and social, political and economic participation on the basis of human individuality, motivation, and opportunity. This is actually an institutionalized compensation for the reverse situation:

> Woman alone gives birth to children, nurses them, and is their primary nurturer in their early formative years. Moreover, the social and economic roles that commonly have been defined as the province of the male have never been performed exclusively by men. Subconsciously, men are aware of this fact.... *The male has never had an exclusive social or economic role that woman could not participate in too. . . .*
> . . . Awareness of woman's monopoly was psychologically repressed and overshadowed by institutionalizing and socially legitimating male values that had the effect of creating self-fulfilling prophecies (emphasis mine).[11]

DISTINCTIONS BETWEEN MEN AND WOMEN

The Qur'an acknowledges the anatomical distinction between male and female. It also acknowledges that members of each gender function in a manner which reflects the well-defined distinctions held by the culture to which those members belong. These distinctions are an important part of how cultures function. For this reason, it would be unwise if the Qur'an failed to acknowledge and, in fact, sympathize with culturally determined, functional distinctions.

> As they are divided, so genders are also interwoven differently in each culture and time. They can rule separate territories and rarely intertwine, or they can be knotted like the lines in the Book of Kells. Sometimes no basket can be plaited, no fire kindled, without the collaboration of two sets of hands. Each culture brings the genders together in its unique way.[12]

The Qur'an does not attempt to annihilate the differences between men and women or to erase the significance of functional gender distinctions which help every society to run smoothly and fulfil its needs. In fact, compatible mutually supportive functional relationships between men and women can be seen as part of the goal of the Qur'an with regard to society.[13] However, the Qur'an does not propose or support a singular role or single definition of a set of roles, exclusively, for each gender across every culture.

The Qur'an acknowledges that men and women function as individuals and in society. However, there is no detailed prescription set on how to function, culturally. Such a specification would be an imposition that would reduce the Qur'an from a universal text to a culturally specific text—a claim that many have erroneously made. What the Qur'an proposes is transcendental in time and space.[14]

Gender distinctions and distinct gender functions contribute to the perceptions of morally appropriate behaviour in a given society. Since the Qur'an is moral guidance, it must relate to the perceptions of morality—no matter how gender-specified—which are held by individuals in various societies. Yet, the mere fact that the Qur'an was revealed in seventh-century Arabia when the Arabs held certain perceptions and misconceptions about women and were involved in certain specific lewd practices against them resulted in some injunctions specific to that culture.

Some prevailing practices were so bad they had to be prohibited explicitly and immediately: infanticide, sexual abuse of slave girls, denial of inheritance to women, *zihar*,[15] to name a few of the most common. Other practices had to be modified: polygamy, unconstrained divorce, conjugal violence, and concubinage, for example. With regard to some practices, the Qur'an seems to have remained neutral: social patriarchy, marital patriarchy, economic hierarchy, the division of labour between males and females within a particular family.

Some women activists today openly question this neutrality. Why didn't the Qur'an just explicitly prohibit these practises? If the evolution of the text and its **overall** objective is consumed under one—albeit important—aspect of social interaction, say consciousness raising with regard to women, then the Qur'an is made subservient to that aspect, rather than the other way around. There is an essential acknowledgement of the relationship between men and women as they function in society, but it is not the sole nor primary objective of the text.

In addition, certain practices encouraged by the Qur'an may be restricted to that society which practised them, but the Qur'an is 'not confined to, or exhausted by, (one) society and its history....'[16] Therefore, each new Islamic society must understand the principles intended by the particulars. Those principles are eternal and can be applied in various social contexts.

For example, in Arabia at the time of the revelation, women of wealthy and powerful tribes were veiled and secluded as an indication of protection. The Qur'an acknowledges the virtue of modesty and demonstrates it through the prevailing practices. The principle of modesty is important—not the veiling and seclusion which were manifestations particular to that context. These were culturally and economically determined demonstrations of modesty.[17] Modesty is not a privilege of the economically advantaged only: all believing women deserve the utmost respect and protection of their modesty—however it is observed in various societies.

Modesty is beneficial for maintaining a certain moral fibre in various cultures and should therefore be maintained—but on the basis of faith: not economics, politics or other forms of access and coercion. This is perhaps why Yusuf Ali translates verse 24:31 'what (must ordinarily) appear'[18] (with regard to uncovered parts), to indicate that (ordinarily) there are culturally determined guidelines for modesty.

This method of restricting the particulars to a specific context, extracting the principles intended by the Qur'an through that particular, and then applying those principles to other particulars in various cultural contexts, forms a major variation from previous exegetical methodologies. The movement from principles to particulars can only be done by the members of whatever particular context a principle is to be applied. Therefore, interpretation of the Qur'an can never be final.

Key Terms and Concepts in the Qur'an

In addition to analysing gender in Qur'anic Arabic, I also analyse certain key words and expressions with regard to humankind in general and to woman in particular, in order to reveal a contextual understanding. Words have a basic meaning—that which can be understood by it, in isolation—and a relational meaning—that connotative meaning derived from the context in which that term is used.[19]

The Qur'an has its own *Weltanschauung*.[20] Despite the basic meaning of certain terms prior to the revelation of the Qur'an, some of the same terms hold different indications specific to their usage in the Qur'an. Each word must be understood within its contextual constraints. As discussed above, I will also demonstrate that there is a distinct relationship between the Qur'an and gender specifications commonly understood as part of Arabic.

What is more, particular emphasis must be made on language used to discuss the Unseen. The Unseen is a part of reality concealed or hidden from human perception. As such, all discussions that the Qur'an contains about matters from the Unseen involve the ineffable: the use of language to discuss what cannot be uttered in language. Such language cannot be interpreted empirically and literally.

Finally, a correlation needs to be made between guidance and every theme discussed in the Qur'an. The Qur'an identifies itself as *hudan*: guidance. When it is shown that guidance extends beyond the normal boundaries which distinguish one human being from another, then it will be seen that it extends beyond gender distinction.

A Word About Quotations and Translations of Qur'an

Throughout this book I have chosen to use Muhammad Marmaduke Pickthall's *The Glorious Qur'an: Text and Translation*, with several alterations. Each Qur'anic usage of the word *insan*, I have translated as 'humankind'. More importantly, I have exchanged archaic terms with more modern ones.

Occasionally, when a verse is discussed at length, the original Arabic text has been transliterated and my own translation has been offered—especially of key terms. This is necessary for the particular interpretation I have made of the text involved.

Finally, there are a few instances in which I have used the translation offered by A. Yusuf Ali in *The Holy Qur'an: Text, Translation and Commentary*. These are marked in the endnotes.

With regard to the Qur'anic quotations in general, they are always noted in the text in parentheses by two numbers with a colon between them, like (1:1-7). The first number refers to the chapter, and the number(s) after the colon refer to the verse(s). In a few instances, I have given the name of the chapter in transliteration, but I have also included the number of this chapter.

Chapter Outlines

This Introduction contains background information and analysis of methodology.

In Chapter 1, I will review some of the problems inherent in a haphazard analysis of the creation of humankind. I propose a cohesive thread in human creation which centres on the pairing essential to all creation. Therefore, both the male and the female are significant in the creation of humanity and have primal equality.

In Chapter 2, I look at the various roles represented by some key Qur'anic female characters and analyse the implications of these to our perceptions of the 'proper' roles and functions of women in Islamic societies. This chapter presents the Qur'an's sociological implications for women. Another dimension of this is taken up in Chapter 4.

Chapter 3 draws relationships between the Qur'anic egalitarian emphasis when discussing the Hereafter and its overall equitable intent. The equity of recompense is meant at one and the same time as an inspiration towards the afterlife and as an explanation of the overall Qur'anic scheme of justice and equity.

In Chapter 4 I make a philosophical analysis of gender in the Qur'an. Through it, I demonstrate the potential of the Qur'an to overcome the oversimplification that has characterized some traditional interpretations which have repressed the potential of woman. As such, the Qur'an can be applied to women within the complexities of developing civilizations. I will criticize some of the long-standing limitations put on or practised against women as a result of such a limiting perspective.

---

[1]The perspectives, circumstances and background of the individual. This concept will be discussed at greater length below.

[2]One notable exception in English is Fazlur Rahman's *Major Themes of The Qur'an* (Chicago and Minneapolis: Bibliotheca Islamica, 1980). In addition, see Mustansir Mir, *Thematic and Structural Coherence in the Qur'an: A Study of Islahi's Concept of Nazm*, (University of Michigan Microfilms International, 1987), which gives a comprehensive analysis of the significance of theme to Qur'anic organization and, consequently, its exegesis.

[3]See Majorie Procter-Smith, *In Her Own Rite: Reconstructing Feminist Liturgical Tradition* (Nashville: Abingdon Press, 1990), Chapter 1, pp. 13-35, on the significance of inclusion and exclusion of women in religious dialogue.

[4]For example, Fatna A. Sabbah in her book *Woman in the Muslim Unconscious* translated by Mary Jo Lakeland from the French *La Femme dans L'inconscient musulman*, (New York: Pergamon Press, 1984) discusses valid points with regard to this issue, but when she discusses the Qur'an, she fails to distinguish between the Qur'an and the Qur'anic interpreters.

[5]For details of Fazlur Rahman's discussion of the above double movement methodology—'from the present situation to Qur'anic times, then back to the present'—for particular communities, see his *Islam and Modernity: Transformation of an Intellectual Tradition* (Chicago: The University of Chicago Press, 1982), Introduction, especially pp. 4-9.

[6]This is a direct contradiction of the classical models which propose that the masculine plural form means male (exclusively). Thus, language is used to make male the norm, and by implication, the female must be abnormal.

[7]This is the Scripture wherein there is no doubt, a guidance unto those who ward off (evil), who believe in the unseen, establish worship, and spend of that We have bestowed upon them.... (Verse 2:2-3).

[8]M.M. Bakhtin, *Speech Genres and Other Late Essays*, trans. Vern W. McGee, (eds.) Caryl Emerson and Michael Holquist (Austin: University of Texas Press, 1986), pp. 6-7.

[9]This recurrent problem in 'reading' the Qur'an causes readers to justify limiting to women statements made using feminine forms and figures; see Chapter 3 below concerning Bilqis. Although she is a good leader—that happens to be a woman—she is not taken as a universal example of leadership. Statements using masculine forms or figures are limited to men and extended to women only by *qiyas* (analogical reasoning).

[10]Alvin J. Schmidt, *Veiled and Silenced: How Culture Shaped Sexist Theology* (Macon, Georgia: Mercer University Press, 1989), Introduction, pp. xiii.

[11]Ibid., pp. 59-60.

[12]Ivan Illich, *Gender* (New York: Pantheon Books, 1982), pp. 106-7.

[13]See Sayyid Qutb, *Fi Zilal al-Qur'an*, 6 vols. (Cairo: Dar al-Shuruq, 1980), Vol. II, pp. 642-3, where he discusses the shared benefits and responsibility between men and women in the Islamic social system of justice.

[14]Fazlur Rahman, *Islam and Modernity*, pp. 5-7, discusses the moral values of the Qur'an in 'extra-historical transcendental' terms, that is, the moral value extracted from a particular verse goes beyond the time and place of the specific instance at which that verse and its injunction was occasioned.

[15]The practice of stating that one's wife was as 'the back of my mother', which would make conjugal relations impossible, but would not totally free the woman for remarriage.

¹⁶Wan Mohd Nor Wan Daud, *The Concept of Knowledge in Islam and Its Implications for Education in a Developing Country* (London: Mansell Publisher Limited, 1989), p. 7.

¹⁷See William Robertson Smith, *Kinship and Marriage in Early Arabia*, (ed.) Stanley A. Cook (London: A. & C. Black, 1907).

¹⁸Translation by A. Yusuf Ali, *The Holy Qur'an: Text, Translation and Commentary*, US ed. (Elmhurst, N.Y.: Tahrike Tarsile Qur'an Inc., 1987).

¹⁹Toshihiko Izutsu, *God and Man in the Koran: Semantics of the Koranic Weltanschauung* (Tokyo: The Keio Institute of Culture and Linguistic Studies, 1964), p.17.

²⁰Ibid, 'world-view' derived from a study of the concepts and key words used in a particular context; in this case, in the Qur'an.

CHAPTER 1

In the Beginning, Man and Woman were Equal:
Human Creation in the Qur'an

*H*OW does the Qur'an describe the creation of woman? Do the Qur'anic accounts of the process of the creation of humankind distinguish woman from man in such a way as to confine her potential to a single biologically determined role? Does it imply created inferiority? Despite the distinctions between the treatment of men and the treatment of women when the Qur'an discusses creation of humankind, I propose that there is no essential difference in the value attributed to women and men. There are no indications, therefore, that women have more or fewer limitations than men.

The Qur'an does not consider woman a type of man in the presentation of its major themes. Man and woman are two categories of the human species given the same or equal consideration and endowed with the same or equal potential. Neither is excluded in the principal purpose of the Book, which is to guide humankind towards recognition of and belief in certain truths. The Qur'an encourages all believers, male and female, to follow their belief with actions, and for this it promises them a great reward. Thus, the Qur'an does not make a distinction between men and women in this creation, in the purpose of the Book, or in the reward it promises.

Creation and the Language of the Unseen

As with other matters of the Unseen, the full details of creation are beyond human language and comprehension. As Kenneth Burke suggests, 'Language is intrinsically unfitted to discuss the supernatural

literally'. Words about God and the Unseen 'must be used ana-
logically' because these matters 'transcend all symbol-systems'.[1] Yet,
all we have with which to discuss these matters is the words of human language—the same words we use to discuss empirical matters.

The Qur'an says, 'He it is Who has revealed unto you (Muhammad) the Scripture wherein are clear revelations. They are the substance of the Book—and others (which are) allegorical' (3:7). The
complete meaning of allegorical verses cannot be empirically determined. Every discussion of the Unseen involves the ineffable.
Eventually it ends upon itself: a discussion about the words used to
discuss that which is unattainable in language. Therefore, my discussion of the creation of woman and man in the Qur'an is primarily a discussion of language.

Creation of the First Parents

All the Qur'anic accounts of the creation of humankind begin with
the original parents: 'Oh children of Adam! Let not Satan seduce
you as he caused your first parents to go forth from the Garden.'
(7:27). We assume that our 'first parents' were like us. Indeed this
assumption is well founded in all but the major consideration of this
chapter: their creation. All human beings after these two were created 'in the wombs of their mothers'. Serious implications have
been drawn from the discussions, myths, and ideas about the creation of the first parents which have had lasting effects on attitudes
concerning men and women.[2]

The Creation of Humankind

According to Maududi,[3] the entire process of human creation was
in three steps: **1.** the initiation of creation; **2.** the formation or perfection; and **3.** the bringing to life. He bases his analysis on verses
like the following:

Just recall the time when your Lord said to the angels, 'I am going to create a
human of clay: when I perfect it in every way, and blow into it of my *ruh*,[4] all of you
should bow down before it.' (38:71-72).[5]

The Qur'an often uses some form of the word *khalaqa*[6] to refer

16

to the first step in the creative process, the initiation of creation. However, *khalaqa* has also been used for the second step, for the creation of each and every human, and also for the creation of everything[7]. Each human is created. Everything is created.

The word *sawwara*—to 'form', 'shape', 'design' or 'perfect'—in this case, to give human specifications, describes the second step in the creative process. 'Perfection' means that Allah formed humankind exactly as He wanted it to be. 'Surely We created humankind in the best stature' (95:4), 'Allah it is Who ... fashioned you, and perfected your shapes' (40:64).

Allah demonstrates in the Qur'an that the form given to humankind is that form best suited to fulfil its vicegerency on earth. In addition, one characteristic of human creation is the two distinct but compatible genders. The two constitute a part of that which 'perfects' the human created form. Thus, the creation of the human form was a conscious decision by Allah—'Who gave everything He created the best form' (32:7).

The third and final step in the creation of humankind is that step which elevates them above the rest of creation: the breathing of the Spirit of Allah (*nafkhat al-ruh*) into each human—male or female.[8]

The Origins of Humankind

Wa min ayatihi an khalaqa-kum min nafsin wahidatin wa khalaqa min-ha zawjaha wa baththa minhuma rijalan kathiran wa nisa'an (4:1).

And *min* His *ayat* (is this:) that He created You (humankind) *min* a single *nafs*, and created *min* (that *nafs*) its *zawj*, and from these two He spread (through the earth) countless men and women. (4:1).

The above verse[9] presents the basic elements in the Qur'anic version of the story of the origins of humankind, that story commonly understood as the creation of Adam and Eve. To aid our understanding, we will look at these four key terms: *ayat, min, nafs* and *zawj*.[10]

AYAH

An *ayah* (pl. *ayat*) is 'a sign' which indicates something beyond itself.

Just as a waymark must not cause a traveller to rivet his eyes on itself, but direct him towards a certain place which is the real destination of his travel, so every phenomenon, instead of absorbing our attention *as a natural phenomenon*, and transfixing it immoveably to itself, should act always in such a way that our attention be directed towards something beyond it.[11]

A tree, in common understanding, is only a tree. In true understanding, a tree is an *ayah* which reflects the presence of Allah. A tree, and other phenomena of nature, are implicit *ayat*: empirical signs which can be perceived by humankind. The known world— the Qur'anic (*'alam al-shahadah*) 'Seen' world—is made of implicit or non-linguistic *ayat* (2:164, 3:190), including our own creation (51:49).

Explicit *ayat* are linguistic: verbal symbols or words. In addition to reiterating the significance of the implicit, non-verbal or conceptual *ayat* of the known world, words of revelation testify to specific information about the realm of the Unseen world (*'alam al-ghayb*). Explicit *ayat* which give information about the Unseen cannot be discovered or fully perceived by ordinary human faculties.[12] They are only known through revelation. The explicit or linguistic *ayat* of revelation are irrevocably linked with creation.[13] In the Qur'an, both linguistic and non-linguistic *ayat* are considered Divine[14] and both are intended to help complete the purpose of the Book, i.e. to provide guidance.

MIN

Min primarily has two functions in Arabic. It can be used for the English preposition 'from' to imply the extraction of a thing from other thing(s). It can also be used to imply 'of the same nature as'. Each usage of *min* in the above verse (4:1) has been interpreted with one or both of these two meanings which yield varying results.

For example, in al-Zamakhshari's commentary, the verse means that humankind was created *in/of the same type as* a single *nafs*, and that the *zawj* of that *nafs* was taken *from* that *nafs*. He uses the Biblical version to substantiate his opinion that the *zawj* was extracted from the *nafs*. In addition, other verses on this subject (7:189, 39:6) state that Allah *ja'ala* from the *nafs* its *zawj*. *Ja'ala*, which means 'to create something from another thing', gives *min* the meaning 'from', i.e. extraction. This meaning of *min* gives rise to the idea that the first created being (taken to be a male person)

18

was complete, perfect and superior. The second created being (a woman) was not his equal, because she was taken out of the whole, and therefore, derivative and less than it.

When *min* is taken to mean 'in/of the same type' for both of its occurrences in this verse, it is often because of the verses in the Qur'an which use the same formula with the plurals of *nafs* (*anfus*) and *zawj* (*azwaj*), for example, 16:82 and 42:11: 'and Allah made (*ja'ala*) *azwaj* for you from (*min*) your *anfus*'; 30:21: 'and among His *ayat* is this: that He created *azwaj* for you from your *anfus*, to live with tranquilly.' These are then interpreted to mean 'your mates are the same type or kind as you are'.

Finally, English translations choose the word 'from' for each of the above occurrences of *min,* thus suggesting another possibility. However, these interpretations of *min* do not clarify the meaning of the verse's most significant terms, *nafs* and *zawj*.

NAFS

The term *nafs* has both a common and a technical usage. Although the common usage of *nafs* translates as 'self', and its plural, *anfus*, as 'selves', it is never used in the Qur'an with reference to any created self other than humankind. As for the technical usage in the Qur'an, *nafs* refers to the common origin of all humankind. Despite the accidental consequence of spreading throughout the earth and forming a variety of nations, tribes and peoples with various languages and of various colours, we all have the same single origin.[15]

Grammatically, *nafs* is feminine, taking the corresponding feminine adjectival and verbal antecedents. Conceptually, *nafs* is neither masculine nor feminine, forming, as it does, an essential part of each being, male or female. For this reason it can (and does) also have masculine antecedents.

The term *nafs*, which later in Islamic philosophy and Sufism came to mean soul as a *substance* separate from the body, in the Qur'an means mostly 'himself' or 'herself' and, in the plural, 'themselves', while in some contexts it means 'person' or the 'inner person', i.e. the living reality of man—but not separate from or exclusive of the body. In fact, it is body with a certain life-and-intelligence center that constitutes the inner identity or personality of man.[16]

In the Qur'anic account of creation, Allah never planned to begin the creation of humankind with a male person; nor does it ever refer to the origins of the human race with Adam.[17] It does not even state that Allah began the creation of humankind with the *nafs* of Adam, the man. This omission is noteworthy because the Qur'anic version of the creation of humankind is not expressed in gender terms.

ZAWJ

The other term of significance in verse 4:1 is *zawj*. As a common term, *zawj* is used in the Qur'an to mean 'mate', 'spouse', or 'group', and its plural, *azwaj*, is used to indicate 'spouses'. This is the term used in refering to the second part in the creation of human-kind, whom we have come to accept as Eve, the female of the original parents. However, grammatically *zawj* is masculine, taking the corresponding masculine adjectival and verbal antecedents. Conceptually, it also is neither masculine nor feminine,[18] and is used in the Qur'an for plants (55:52) and animals (11:40), in addition to humans.

We know even less about the creation of this *zawj* than we know about the creation of the original *nafs*. The Qur'an states only two things about its creation: that it is *min* the first *nafs*, and is *zawj* in relation to that *nafs* (4:1, 7:189, 39:6). It is perhaps this scarcity of detail that has caused Qur'anic commentators, like al-Zamakhshari, and other Muslim scholars to rely on Biblical accounts which state that Eve was extracted from (*min*) the rib or side of Adam.[19]

The absence of detail in the Qur'an indicates one or more of the following: **1.** the reader already has enough details about a story to understand it and other details are unnecessary—even redundant; **2.** these details are unimportant to the point which the Qur'an is making at the particular time; **3.** the Qur'an is referring to some-thing Unseen, for which human language is already deficient. Keeping these three reasons in mind, I reiterate that the Qur'an gives very little information about this primal *zawj*.

The Dualism of the Creation

I am interested in the Qur'anic use of *zawj* as one in a necessary or contingent 'pair' essential to the Qur'anic accounts of creation:[20]

20

everything in creation is paired. 'And of all things We have created (*zawjayn*) pairs, perhaps you [will all] reflect [on this fact].' (51:49). Dualism becomes a necessary characteristic of created things.[21]

In this usage, a pair is made of two co-existing forms of a single reality, with some distinctions in nature, characteristics and functions, but two congruent parts formed to fit together as a whole. 'Each member of the pair presupposes the other semantically and stands on the very basis of this correlation.'[22] A man is only a 'husband' in reference to a 'wife'. The existence of one in such a pair is contingent upon the other in our known world. These are the Qur'anic pairs. Night flows into day; the male is irrevocably linked with the female[23] as man is compatibly linked with woman.

With regard to creation, 'everything created in pairs' means that the counterpart of each created thing is part of the *plan* of that thing. 'Glory be to Him Who created all the *azwaj*, of that which the earth grows, and of their own *nafs*, and of that which they know not.' (36:36). Each created thing is contingent upon its *zawj*. In this contingency, the creation of both the original parents is irrevocably and primordially linked; thus, the two are equally essential.

There is ample Qur'anic support for the contention that these pairs in creation are equally essential: 'And all things we have created by pairs (*zawjayn*)' (51:49). 'He Who created all the *azwaj*' (43:12). 'Glory be to (Him) Who created all the *azwaj*: of that which the earth grows, and of their own [human] *anfus*, and *of that which they know not!*' (36:36). The Qur'an first establishes that all created things are paired, then reinforces this mutual necessity by depicting theoretical pairs in the rest of creation.[24]

Although the Qur'an establishes that humankind was intentionally created in the male/female pair—'Allah created you from dust, then from a little fluid [the sperm-drop], then He made you pairs' (35:11), and 'Verily He has created the (two) spouses (*zawjayn*): male and female' (53:45)—and distinguishes between them—'the male is not like female' (3:36)—it does not attribute explicit characteristics to either one or the other, exclusively.

It is obvious that the child-bearing function belongs with the female. 'Allah created you from dust, then from a little fluid [the sperm-drop], then He made you pairs. No female bears or brings forth except with His knowledge.' (35:11-12). 'He It is Who did create you from/of a single *nafs*, and from/of it did make its *zawj*, so that he might take rest in her. And when he [a man] covered her (a woman), she bore a light burden' (7:189).[25] 'Allah knows what every female bears in her womb' (13:8). Although the Qur'an illustrates explicitly the correlation between the female and bearing children, all other functions connected with child care and rearing, if mentioned at all in the Qur'an,[26] are never described as essential created characteristics of the female. Thus, the Qur'anic reference is restricted to the biological function of the mother—not the psychological and cultural perceptions of 'mothering'.

Femininity and masculinity are not created characteristics imprinted into the very primordial nature of female and male persons, neither are they concepts the Qur'an discusses or alludes to. They are defined characteristics applied to female and male persons respectively on the basis of culturally determined factors of how each gender should function. They have figured very strongly in interpretation of the Qur'an without explicit Qur'anic substantiation of their implications.[27]

In the Qur'an, the essential contingent male/female pairs in humankind function on a physical, social and moral level. Just as the essential male/female is contingent, so, too, are the physical beings; there is a tranquil link between the human pair, man and woman: 'Among His signs is this: that He created *azwaj* for you from your own *anfus* so that you may find rest in them' (30:21). Man is intended as a comfort to woman; woman is intended as a comfort to man. This statement does not make it a reality. However, the Qur'an clearly depicts a necessary link between the functional members of each gender, like an echo of the contingency between the essential pairs of all created things.[28]

In conclusion, the following verse depicts the structure of the created social order:

O Mankind! Be careful of your duty to your Lord Who created you from a single *nafs*, and from it created its *zawj*, and from that pair spread abroad [over the earth] a multitude of men and women.' (4:1).

It establishes the origin of all humankind as a single *nafs*, which is part of a contingent-pair system: that *nafs* and its *zawj*. In practical terms, this essential pair is man and woman. In this verse, the use of the words 'men' and 'women' means that the physical manifestations of the essential paired reality are multiplied and 'spread abroad [over the earth]'. The earth is inhabited by many peoples, nations, and cultures. This verse transcends not only time but space as well.

The Events in the Garden

One other gender consideration with regard to the creation of humankind centres on the revelation about the Garden of Eden. The original parents are the two essential characters in the story of the forbidden tree. The third character, Satan, is significant in his interactions with these two. Through this scenario, the Qur'an demonstrates the following concepts: **1**. fundamental guidance in the Qur'anic scheme; **2**. temptation and deception which hinder humankind's efforts; **3**. divine forgiveness; and finally, **4**. individual responsibility. Therefore these events must be reviewed in relation to these concepts.

It is clear from Qur'anic descriptions that the Garden was never intended as the dwelling place of the human species. Part of Allah's original plan in the creation of humankind was for man to function as a *khalifah* (trustee) on earth.[29] In the Garden, humankind had no need to struggle for the basic necessities of life: food, clothing, and shelter. 'It is (vouchsafed) unto you that you hunger not therein, neither are you naked, And you thirst not therein nor are you exposed to the sun's heat.' (20:118-19).

However, in the Garden, and on earth, humankind share the same test: the choice between obedience and disobedience. Allah warns Adam and Eve against approaching one of the trees in the Garden. The Qur'an does not give special attributes to the tree itself: it is merely a symbol of the test.

After the creation of Adam and Eve, Satan shows his true nature by not bowing when ordered to by Allah. He is arrogant and wilfully disobedient. At that time, the original parents—and in effect humankind[30]—are warned against Satan. 'He is an enemy to you' (7:22; 20:117; 12:5; and others). Failure to remain cognizant of this fact can cause one to fail at the test of obedience and ultimately can result in following Satan to the chastisement of Hell.

When Satan approaches the original parents, 'he said: "Your Lord forbade you from this tree only lest you should become angels or become of the immortals." And he swore to them (saying): 'Lo! I am a sincere advisor to you [both].' Thus did he lead them on with guile' (7:21). In the Qur'an, the nature of Satan's temptation of the original parents remains important: when Satan tempts you, he comes disguised as your true friend and might even suggest to you something for your own good. In fact, the offer he made to Adam and Eve was so great that they failed to remember the warning against him that they had been given. They failed to remember Allah's admonition and approached the tree.

Upon recognition of the error that they had made, the original parents repented and asked for forgiveness (7:23). Their Lord not only accepted their repentance and forgave them, He demonstrated a very special feature of Himself: mercy and grace. He extended to them, and to humankind at large, the explicit guidance—revelation. This story ends with this moral: any human might disobey through forgetfulness, the general nature of human weakness, and the temptations of Satan, but he who recognizes his error, repents, and asks for forgiveness, can and will be forgiven.

Moreover, guidance is always available to humankind to remind them of their commitment to Allah and the guile of Satan, the enemy. This is a special mercy from their Lord. However, whoever disobeys through arrogance and intentional rebellion has been promised due punishment and eternal damnation. He is like Satan, who disobeyed and persisted in his arrogant, disobedient ways.

The story emphasizes that human beings are susceptible to Satan's temptation. Under such temptation, they are prone to forget the agreements they have made with the Creator (whether a specific agreement as in this case, to stay away from the tree, or a general agreement, as with all humankind, to remain servants true to the cause of righteousness), and as a consequence of forgetting, to disobey. The story reminds humankind in explicit terms that such forgetting can be forgiven.

It is noteworthy that, with one exception, the Qur'an always uses the Arabic dual form to tell how Satan tempted both Adam and Eve and how they both disobeyed. In maintaining the dual form, the Qur'an overcomes the negative Greco-Roman and Biblical–Judaic implications that woman was the cause of evil and damnation.[31] Moreover, it signifies the Qur'anic emphasis on the individual responsibility: '... God does not change the situation of a people until they change it themselves [anfus]' (13:11, 8:53), i.e. unless humans, individually and collectively, take the initiative, there can be no change for better or worse. 'The Qur'an states repeatedly that every man and woman individually and every people collectively are alone responsible for what they do.'[32]

The one exception to the Qur'anic use of the dual form to refer to the temptation and disobedience of Adam and Eve in the Garden singles out Adam:

And verily We made a covenant of old with Adam, but he forgot, and We found no constancy in him. . . . And the devil whispered to him saying: 'Oh Adam! Shall I show you the tree of immortality and power that does not waste away?' Then the two of them (Adam and his wife) ate of the fruit (of the forbidden tree)... And Adam disobeyed his Lord, so went astray (20: 115-21).

This passage comes after some verses which refer to the impatience of the Prophet for the Qur'anic revelation. The Prophet used to try to memorize the verses as they were revealed for fear of forgetting. However, the Prophet need not have worried. Here, in the story of Adam, the point of forgetfulness is mentioned. It is Satan who seduces man to forget. Allah forgives Adam, accepts his repentance and gives him guidance. This reveals Allah's mercy and His guardianship over his servants and over the guidance. Adam is isolated in these verses because of a particular point that is being made. This is an example of the omission of details in the Qur'an. Nevertheless, this much is clear: woman is never singled out as the initiator or temptress of evil.

Conclusion

The Qur'anic account of the creation of humankind relates to other themes in the overall Qur'anic *Weltanschauung*: *tawhid*, guidance, individual moral responsibility and equality. For example, the phenomenon of pairs in creation supports the major Qur'anic principle of *tawhid*: the unicity of Allah. The Qur'an states explicitly that 'nothing is like Him' (42:11). Philosophically, since all created things are paired, He who is not created is not paired: the Creator is One.

The Qur'anic version of human creation establishes a special link between the Creator, Allah, and the created, humankind. That link is the basis for the existence of the Qur'an and for the guidance which is connected to the creation. At the moment Adam came to earth, the basis of the relationship established between the Creator and His created human beings was completed through guidance or revelation. 'He said, "Go down hence (Satan and humankind), both of you, one of you a foe unto the other. But if there come unto you from Me a guidance, then whoever follows My guidance, he will not go astray nor come to grief."' (20:123).

The unique and dynamic relationship between the Creator and His creature is also represented in the *ruh* of Allah which is blown into each being, male and female. Both the *ruh* and the guidance aid in the struggle to pass the test on earth, to resist Satan's temptation and to conclude in an eternal happiness.

Although the male and female are essential contingent characters in the creation of humankind, no specific cultural functions or roles are defined at the moment of creation. At that moment, Allah defines certain traits universal to all humans and not specific to one particular gender nor to any particular people from any particular place or time. The divine *ayat*, in both their words of revelation and empirical forms in nature, are available to all. The empirical *ayat* can be perceived by every person anywhere and at any time. The specific *ayat* which Allah has revealed to a chosen few at particular times under particular circumstances are meant for all.

The Qur'anic version of the story of the Garden signifies individual responsibility. The *nafs* represents that individuality. Therefore, whatever good is performed is reflected on to that *nafs* and whatever bad is performed has its consequences on that *nafs*.[33]

The Qur'anic account of the creation of humankind is important, above all, because it points out that all humans share a single point of origin. That point is represented in the Qur'anic accounts of creation by its use of the term *nafs*. Just as we have one point of origin,

26

so do we also have one destination: from one to many and back to one again. What remains to be seen then is the Qur'anic treatment of the dynamics involved in the interactions between the many.

---

[1] Kenneth Burke, *The Rhetoric of Religion* (Boston: Beacon Press, 1961), p. 14, footnote.

[2] Most important of which is the assumption that human creation began with a man: which gives all men *a priori* superiority over all women.

[3] S. Abul A'la Maududi, *The Meaning of the Qur'an*, edited by A.A. Kamal and translated by Ch. Muhammad Akbar, 6th edn., 13 vols. (Lahore, Pakistan: Islamic Publication Ltd., 1983), Vol. 4, p. 11.

[4] Pl. *arwah*, loosely translated as 'spirit', will be discussed below in greater detail.

[5] Translation taken from his *tafsir*.

[6] *Khalaqa*: to create, to bring something into existence from a state of non-existence.

[7] Verse 25:2—'Who created each and everything and then ordained its destiny.'

[8] This is the closest that the Qur'anic version of creation comes to the Biblical version of man in 'God's image'.

[9] And others like it with similar wording and only a few slight changes. See verses 7:189 and 39:6. See also verse 6:98— *'ansha'na-kum min nafsin wahidatin*: We formed you (all) from a single *nafs.'*

[10] *Ayah* pl. *ayat*, *nafs* pl. *anfus* and *nufus*, *zawj* pl. *azwaj*. These words will be transliterated throughout the text because of their distinctive untranslatable quality. They will all be discussed in detail in this section.

[11] Toshihiko Izutsu, *God and Man in the Koran: Semantics of the Koranic Weltanschauung* (Tokyo: The Krio Institute of Culture and Linguistic Studies, 1964), p. 134.

[12] This is why early Muslim philosophers distinguished revealed knowledge from empirical knowledge.

[13] See verses 87:3, 20:50, and 7:29-30. More importantly, Adam becomes the first Prophet, carrier of the explicit *ayat*. Guidance is promised to humankind and when it becomes lost, obscured, or corrupted over time, it is revitalized. Hence, with Adam begins the tradition of prophecy which continues until Muhammad, who is given the explicit *ayat* in the form of the Qur'an, which remains intact as the legacy of revelation for all who come after.

[14] There are numerous verses which start 'And among His *ayat* ...' and proceed to mention known phenomena of good and bad merit as well as some Unknown phenomena.

[15] Verses 4:1 and 49:13.

[16] Fazlur Rahman, *Major Themes in the Qur'an* (Chicago and Minneapolis: Bibliotheca Islamica, 1980), p. 112.

[17]See also Muhammad Ahmad Khalaf-Allah, *Al-Fann al-Qasasi fi al-Qur'an al-Karim* (Cairo: Maktab al-Anjali Masriyyah, 1965), p. 185, where he discusses the Qur'an's reasons for not using the term 'Adam' to discuss the origins of humanity.

[18]See verses 4:20 and 2:102, where *zawj* is used to indicate females, and 2:230 and 58:1, where it is used to indicate males.

[19]Not to be overlooked, however, is a similar story in *ahadith*—all of which have a single link in their *isnad*, thus diminishing their strength. See Riffat Hassan, 'Made from Adam's Rib: The Woman's Creation Question', *Al-Mushir Theological Journal of the Christian Study Centre, Rawalpindi, Pakistan*, Autumn 1985, pp. 124-56 for a detailed analysis of these.

[20]It is also important to the Qur'anic account of the Hereafter and will be discussed in Chapter 3.

[21]See Qutb, Vol. 2, p. 648.

[22]Izutsu, *God and Man*, p. 85.

[23]'And that He created the two spouses (*zawjayn*), the male and the female.'(53:45).

[24]For example, 'Glory be to Him Who created all the sexual pairs, of that which the earth grows, and of themselves, and of that which they know not!' (36:36).

[25]See also verses 13:8, 31:14, and 41:47 for other references to females and procreation.

[26]This is discussed at greater length below.

[27]See discussions below on how the actions of particular individuals have been interpreted as feminine, and therefore exclusively for females, especially when I discuss the Queen of Sheba and the two women of Madyan.

[28]Qutb, Vol. 2, pp. 618-19, discusses the necessary functional link between man and woman.

[29]See verses 2:30 and 38:26, where Allah speaks to the angels and informs them that He plans to create a *khalifah* (trustee) on the *earth*.

[30]The significance of Adam as a metaphorical representative of humankind stems from verses 38:71-2. The creation of the first *nafs* establishes the existence of the entire human race. Each being after Adam must have a *nafs*. Thus Adam was created as the basic human. Much of what we must understand about human qualities the Qur'an demonstrates in a rudimentary form with Adam: creation, trusteeship, test of this world, the role of Satan, guidance, obedience to Allah.

[31]For an excellent review of this dilemma, see Alvin J. Schmidt, *Veiled and Silenced*, Chapter 3: Woman as 'Evil', pp. 39-68.

[32]Rahman, *Major Themes in the Qur'an*, p. 19.

[33]As is continually repeated in the Qur'an, for example: verses 10:108, 27:40, 29:6, 31:12, 35:18, 41:47, and 48:10, among others. When I discuss the equity of recompense in Chapter 3, I will also review this in greater detail.

CHAPTER 2

The Qur'anic View of Woman in This World

A thorough study of the Qur'anic treatment of the subject of woman must include a review of the roles fulfilled by the female characters mentioned, whether explicitly or indirectly, in the text. What is the significance of these women in the circumstance in which they have been mentioned? Are they mentioned for women in particular or for all readers, to fulfil the Qur'anic purpose of guidance? Does the Qur'an propose a single uniform role that women should play in society?

I propose that the Qur'an does not support a specific and stereotyped role for its characters, male or female. The roles of the women who have been referred to in the Qur'an fall into one of three categories: **1**. A role which represents the social, cultural, and historical context in which that individual woman lived—without compliment or critique from the text. **2**. A role which fulfils a universally accepted (i.e. nurturing or caretaking) female function, to which exceptions can be made—and have been made even in the Qur'an itself. Finally, **3**. a role which fulfils a non-gender specific function, i.e. the role represents human endeavours on the earth and is cited in the Qur'an to demonstrate this specific function and not the gender of the performer, who happens to be a woman.[1]

How the Qur'an Teaches the Reader through the Events in the Lives of the Individuals It has Mentioned

The Qur'an is a moral history. It proposes moral values, which are 'extrahistorical' and 'transcendental' in nature, such that 'their location at a point in history does not exhaust their practical impact or, one might say their meaning'.[2] The time-span for the Qur'an is not limited to this present world. It is not limited to observable facts and events. It includes not only information about the actual events which occurred, but also about the intent behind such events and about their psychological effect. We have no sure indication for many accounts to determine if they are historical or metaphorical, literal or allegorical.

There are two significant concerns in our look at female characters mentioned in the Qur'an: the Makkan and Madinan chronological periods of revelation, and the references to known historical events. It has been argued that the detailed discussions during the Madinan period indicate accommodation towards the existing community and in some instances limit the universal objectives of the Qur'an. To overcome such a limitation, Muslim jurists and thinkers should give precedence to the universal revelations from the Makkan period.[3]

A hermeneutical model which derives basic ethical principles for further developments and legal considerations by giving precedence to general statements rather than particulars could solve many problems in application. However, there is no indication that all general principles were introduced in the Makkan period and all Madinan revelations were specific. Thus, although an analysis is needed which gives precedence to general principles over specific statements, it will not necessarily follow that this will be only a chronological consideration of the difference between Makkan and Madinan passages.

As for references in the Qur'an to well-known historical events, a significant part of Qur'anic exegesis is knowledge of *asbab al-nuzul*, often translated as the 'reason(s) for revealing (a particular verse or verses)', should more accurately be the 'occasions upon which (or, sometimes, because of which) a certain verse was revealed'. This distinction broadens the application of underlying principles.

If the 'reason' for a verse is a specific incident, then the universal 'extrahistorical' nature of the Qur'an is removed. It becomes a history book in the limited sense of a record of events. However, remembering that the Qur'an was revealed at a particular time in history and within a certain social context helps one to understand the significance of some verses, without implying in any way a restriction of the Qur'anic principles to that context.

The Qur'an is quite selective about historical details which help to fulfil its purpose of universal guidance. It promotes the application of the precepts exemplified by those characters mentioned in the verses rather than merely gives a record of the events themselves. For if the reader assumes he is reading a record, what he reads may be insignificant to his own life. The Qur'an gives just enough—but not too much—details in order to facilitate its purpose.

The Qur'an does not distract the reader from the moral principles by overburdening him with details about particular individuals. Thus, a process of interpretation that gets at the significance of the references to particular characters needs to be applied. Fazlur Rahman proposes that 'the process of interpretation . . . consists of a double movement from the present situation to the Qur'anic times, then back to the present' because the

Qur'an is the divine response, through the Prophet's mind, to the moral-social situation of the Prophet's Arabia,... The Qur'an and the genesis of the Islamic community occurred in the light of history and against a social-historical background. The Qur'an is a response to that situation, and for the most part it consists of moral, religious, and social pronouncements that respond to specific problems confronted in concrete historical situations....[4]

In particular, he defines that double movement of interpretation to be from the specific—a study of the historical situation of the problem which the statement addresses, in order to understand its import or meaning—to the general—'generalize the specific answers and enunciate them as statements of general moral-social objectives that can be "distilled" from specific text in the light of the socio-historical background.'[5]

The Significance of the Women Mentioned or Referred to in the Qur'an

I have tried to demonstrate the significance of particular female characters mentioned in the Qur'an to the modern context, because 'if the results of understanding fail in application now, then

31

either there has been a failure to assess the present situation correctly, or a failure in understanding the Qur'an.'[6] To do so, however, I first demonstrate the significance of these women in the context of the text.

I have divided all of the women—whether mentioned explicitly or referred to in passing—into two major categories.[7] Those with no independent significance I have placed into category 1. This category I have divided into two levels: those with very slight mention and those women who performed functions characteristic of the 'proper' role for a woman. At both levels the women are inconsequential to the overall Qur'anic purpose of guidance, but are necessary for the coherence of a story or event. At the second level of this category, the women functioned in accordance to the social constraints of their particular circumstance. The Qur'an does not portray them as universal 'examples'.

Qur'anic examples provide concrete information about the application of moral precepts. In order to apply the moral principles of the Qur'an, a reader must have some practical understanding. Practices are rooted in contexts. Readers who interpret the significance of the women cited in the Qur'an often come to the text with notions of appropriate functions for women. When these are supported on the surface of the Qur'anic portrayal, they do not look further at the examples. This has led to a great deal of oversimplifications and contradictions when the perspective of the individual exegete is superimposed on to the Qur'an itself.[8]

The Qur'an is not a manual of directives which only commands the individual reader to perform certain actions or fulfil particular characteristics. By citing concrete events, it makes conceptual ideas tangible. The female and male characters are particularly important to demonstrate certain ideas about guidance. The characters and events in the Qur'an should always be examined in the light of this overall goal.

It should be noted that all references to female characters in the Qur'an use an important cultural idiosyncrasy which demonstrates respect for women.[9] Except for Mary, the mother of Jesus, they are never called by name. Most are wives and the Qur'an refers to them by means of a possessive construction (the *idafah*) containing one of the Arabic words for wife: *imra'ah* (woman), *nisa'* (women), or *zawj* (spouse, or mate) pl. *azwaj*, and the name of a particular male; for example, the *imra'ah* of Imran, or the *zawj* of Adam.

Even an unmarried woman or one whose husband is not mentioned is linked with some male: Ukht-Musa, the sister of Moses; Ukht-Harun, sister of Aaron, another name for Mary; Umm-Musa; the mother of Moses. However, this particular manifestation of respect is restricted to that context. The general principle—that women should be addressed respectfully—is intended for those who read the Qur'an at other times.

The Qur'an presents examples through several methods. A specific event is spoken of in general terms and the general application of the principles which underlie the verse is emphasized. 'The Qur'anic *ayah* is intended as a continual reality independent of the individuals (mentioned), and the individuals are merely examples of this reality.'[10]

Other verses include information about very specific incidents, but the moral lesson to be drawn from the incidents can be generally applied once they have been interpreted. 'The method the Qur'an uses in this type of incident is to describe the activity which occurred and to describe with that activity, the apparent and hidden legalities (to be deduced from that incident).'[11]

In *Surat-al-Tahrim* (66), the Qur'an points directly to some women as models:

Allah cites an example for those who disbelieve: the wife of Noah and the wife of Lot, who were (respectively) under two of our righteous servants, yet they were betrayed them so that they (the husbands) were of no avail to them against Allah and it was said (to them): 'Enter the fire along with those who enter!'

And Allah sets forth an example to those who believe: the wife of Pharaoh when she said 'My Lord! Build for me a home with You in the garden, and deliver me from Pharaoh and his work, and deliver me from evil doing folk.

And Mary, daughter of Imran whose body was chaste, therefore We breathed therein something of Our *ruh*. She put faith in the words of her Lord and His scriptures and was among the *qanitin* (66:10-12).

These 'examples' which Allah specifically 'cites' for the readers are usually interpreted as being exclusively for women. Yet the verse

33

identifies them as examples for 'those who believe' and 'those who disbelieve'. In other words, they are non-gender specific examples—in this case, of the individual responsibility towards belief. No one is helped to salvation because of an affiliation with someone who strives for good—no matter how intimate that affiliation.[12] Each person must strive for good, or, at the very least, as with the wives of 'righteous servants', must not strive against the truth, and be condemned to the Fire.

Finally, on rare occasions, some incidents mentioned in the Qur'an have exclusive application. There is some historical information which can be extracted from them with very little or no significant application beyond that incident. These are Madinan verses, and the information includes something exclusive to the Prophet Muhammad and his household.[13]

Woman as an Individual

In my discussion of woman in the Qur'an, I have considered the difference between woman as an individual and woman as a member of society. For the most part, the Qur'anic consideration of woman on earth centres on her relationship to the group, i.e. as a member of a social system. However, it is also important to understand how the Qur'an focuses on woman as an individual because the Qur'an treats the individual, whether male or female, in exactly the same manner: that is, whatever the Qur'an says about the relationship between Allah and the individual is not in gender terms. With regard to spirituality, there are no rights of woman distinct from rights of man.

With reference to the individual, the Qur'an most often uses the term *nafs*. On earth, the individual is given responsibility and capacity. Both determine the recompense of the individual in the Hereafter.[14]

Individual capacity is expressed as follows: 'Allah does not tax a *nafs* beyond its scope. For it (is only) that which it has earned, and against it (is only) that which it has deserved' (2:286). There is no distinction between the male and the female with regard to individual capacity. With regard to their potential relationship with

34

Allah, they are the same. With regard to personal aspirations, they are also the same. This is important because:

In every society, in every century, people have assumed that males and females are different not merely in basic anatomy, but in elusive qualities of spirit, soul and ability. *They are not supposed to do the same things, think the same way, or share the same dreams and desires* [emphasis mine].[15]

When various social systems determine differences between men and women, they conclude these differences as indications of different values as well. There is no indication that the Qur'an intends for us to understand that there is a primordial distinction between males and females with regard to spiritual potential. Therefore, whatever differences existing between males and females could not indicate an inherent value, or else free will would be meaningless. The problem arises in trying to determine when and how these differences come into being.

Sayyid Qutb[16] says that 'the *fitrah* [primordial nature] makes the man a man, and the woman a woman', but he goes on to emphasize that this distinction has no inherent value. Al-Zamakhshari,[17] on the other hand, says that men are 'preferred' by Allah over women in terms of 'intelligence, physical constitution, determination and physical strength', although he cites no place in the text which states this. Such an assertion cannot be erased by saying that 'men have no right to overcome women by coercion, or display arrogant behaviour towards them'. Al-'Aqqad[18] says that men deserve preference over women.

I hope to demonstrate the negative effects of interpretations which place an inherent distinction between males and females and then give values to those distinctions. Such interpretations assume that men represent the norm and are therefore fully human. Women, by implication, are less human than men. They are limited and therefore of less value. Such interpretations encourage the stereotypes about women and men which severely hamper the potential of each. In addition, these interpretations justify the restrictions placed on the woman's right to pursue personal happiness within the context of Islam. Most troubling is the tendency to attribute these interpretations to the Qur'an itself rather than to the authors who hold them.

I do not hold such views, nor do I find support for them in the Qur'an. It is interesting to note that even those Muslim authors who issue these interpretations accept that the Qur'an aims to establish social justice. However, it is obvious that their interpretation of social justice does not extend fully to women. It is like Thomas Jefferson and the writers of the American Constitution saying that 'All men are created equal' without intending in the least to include equality between black men and white men.

I propose that the Qur'an depicts human individuals as having inherently equal value by looking at three stages in human existence. First, in the creation of humans, the Qur'an emphasizes the single origin of all humankind: 'He created you (all) from a single *nafs*' (4:1). Second, with regard to development here on earth, the Qur'an emphasizes that the potential for change, growth and development lies within the *nafs* of the individual (or the group) as well: 'Allah does not change the condition of a folk until they (first) change what is in their *anfus*' (13:11). Finally, all human activity is given recompense on the basis of what the individual earns (4:124).[19]

This leaves me to ask the following questions: Is there any way to understand the distinction between individuals? Do these distinctions reflect a particular value system? Do they follow clearly delineated lines between the two sexes?

Distinctions between Individuals: Taqwa

The Qur'an does make distinctions between things and between people. It establishes that the Hereafter is of greater value than this world.[20] It also distinguishes within the Hereafter and on this earth. The value of the distinctions between humankind on earth can be clearly summed up by the Qur'anic statement in *surat Al-Hujarat* (49:13): 'We created you male and female and have made you nations and tribes that you may know one another. *Inna akrama-kum 'inda Allah atqa-kum* [Indeed the most noble of you from Allah's perspective is whoever (he or she) has the most *taqwa*].'

This term *taqwa*, one of the most essential in the Qur'anic *Weltanschauung*, has various translations and definitions.[21] For this research I will consider it to mean 'piety', that is, a pious manner of behaviour which observes constraints appropriate to a social–moral system; and 'consciousness of Allah', that is, observing that manner of behaviour because of one's reverence towards Allah. In the Qur'anic *Weltanschauung*, this term always reflects both the action and the attitude. I do wish to reiterate that this multidimensional term is essential in the Qur'an.

The Qur'anic passage above reconstructs all the dimensions of human existence. It begins with creation. Then, it acknowledges the pair: male and female.[22] These are then incorporated into larger and smaller groups, here translated as 'nation 'and 'tribe' respectively[23] 'That you may know one another', or simply that being distinguishable allows identification. If we were all alike, with nothing to distinguish us, we would have no way of knowing each other or being known.

The culmination of this verse and its central aspect for this discussion is: 'the most noble of you from Allah's perspective is whoever (he or she) has the most *taqwa*.' The distinguishing value from Allah's perspective is *taqwa*. Provided that *taqwa* is understood in both its action and attitude dimensions, this verse is self-explanatory. Allah does not distinguish on the basis of wealth, nationality, sex, or historical context, but on the basis of *taqwa*. It is from this perspective then that all distinctions between woman and woman, between man and man, and between woman and man, must be analysed.

It is worth noting here that this verse follows verses which reprimand the individuals of both genders for mocking, defaming and backbiting one another (49:11-12). One might attribute greater or lesser value to another on the basis of gender, wealth, nationality, religion or race, but from Allah's perspective, these do not form a valuable basis for distinction between individuals (or groups)—and His is the true perspective.[24]

Finally, I wish to point out that the three commentators whose exegetical works were consulted for this research all state that the apparent and worldly things which humans use to judge between one and another are not the true criteria for judgement. They do not agree, however, on what is included in these apparent things. In analysing this verse, only Sayyid Qutb[25] acknowledges that gender is used as a basis for mockery and defamation, which then must be denounced as a false, worldly aspect of superiority. He states that the verses are inclusive of all the variations among humankind: gender, colour, etc., 'because all will return to single scale, that of *taqwa*'.

Al-Zamakhshari[26] mentions that the references to both males and females, with regard to mockery, is more emphatic than the omission of both would be, but does not go on to conclude explicitly that gender could be used erroneously for distinction or superiority.

Maududi[27] says that in these verses 'The whole of mankind is addressed' in order to prevent a great evil which is universally disruptive, and that is prejudice due to 'race, colour, language, country and nationality'. He does not include gender. He quotes Ibn-al-Majah as saying, 'Allah does not see your outward appearances and your possessions, but He sees your heart and your deeds.'[28] I believe that the heart and deeds are genderless on the basis of this verse: 'Whoever does good, from male or female, and is a believer, all such will enter Paradise' (4:124).[29]

Distinctive Female Characters in the Qur'an

I will discuss only three Qur'anic female characters: the mother of Moses, Mary, and Bilqis, the Queen of Sheba. The details in the lives of these women are very different but their stories have been retold by exegetes without the female vision and without focusing on that which transcends their femaleness.

Moses was born at a time when Pharaoh was slaughtering all male Jewish infants. Allah decrees that this child will fill the office of prophecy.

And We inspired the mother of Moses, saying: Suckle him and, when you fear for him, then cast him into the river and do not fear nor grieve. Lo! We shall bring him back to you and shall make him (one) of Our messengers. (28:7)

The first pronouncement in this verse speaks soft words to the mother of Moses. It promises to return the suckling child to her. Then it announces that Allah will make him a messenger. Note the tenderness the Qur'an demonstrates towards this woman's desire to nurture her child. Although the child is living under the threat of Pharaoh's orders to slaughter such children, the child is saved to fulfil the decree of Allah. The desire of his mother towards the child is not directly part of that decree. Yet, the Qur'an details that aspect—in fact, mentions it first.

Returning the child so that he might suckle at the mother's breast will fulfil her maternal need, and presumably make it easy for her to remain silent about the child. Here sacrifice and struggle to fulfil Allah's commandments are tempered with sympathy and tenderness. She is not just a believer, she is a believer with love and concerns, fears and anxieties.

Not to be underestimated in its significance is that the Qur'an states that Umm-Musa received *wahy*,[30] thus demonstrating that women as well as men have been recipients of *wahy*. Thus Umm-Musa demonstrates the point that women are distinct with regard to some aspects, but universal with regard to others.

As I mentioned before, Mary is the only woman referred to by name in the Qur'an. This is because in the Qur'an, Jesus was created in the womb of Mary by special decree and not by normal biology. He is called 'the son of Mary' to demonstrate the significance of his birth. Family relations in the Qur'an are labelled in terms of the father or a male ancestor,[31] but since no father is mentioned for Jesus, he cannot be called in this manner. Most important, from the perspective of the Qur'an, he was not the son of God. Although Mary is called by name in the Qur'an, she is also given one of the honorific titles common at that time in Arabia and employed in the Qur'an when specific women are mentioned: she is called 'sister of Aaron'.

In her story, a messenger appears before her carrying Allah's message that she is to bear a child. She responds, 'How can I have a child and no mortal has touched me, neither have I been unchaste?' (19:20). Once she has accepted the decree, the Divine *ruh* is blown into her, and she bears the child.

When the time for delivery comes, the Qur'an describes her pains of labour and her statement: 'Would that I had died before this time and been long forgotten (rather than to feel such pains).' (19:23). She is like every other woman who bears a child. Then the Qur'an demonstrates Allah's sympathy for her predicament: 'Grieve not!' (19:24) and she is asked to eat, drink, and be comforted (19:26).

Despite the centrality of Jesus to Christianity, no similar affirmation of the unique experience of childbirth is given such detailed consideration in any Christian theological work—not even in the Bible. That special function is elevated to the status worthy of detailed mention to attest to its significance in the Qur'anic worldview. We are not left to just take it for granted.

Again, in this instance, as with Umm-Musa cited above, it is generally considered necessary only that a certain child be delivered to birth or to safety. Yet, the Qur'an includes these details to highlight the fact that the priority of saving the child, in each case, is also viewed in the light of the concern for the respective mothers.[32]

Finally, one simple grammatical consideration with regard to the significance of Mary to all believers. The Qur'an classifies her as 'one of the *qanitin*'[33] (66:12), using the masculine plural form of the word that indicates one devout before Allah. There is no reason not to use the feminine plural form—except to emphasize that the significance of Mary's example is for all who believe—whether male or female. Her virtue was not confined by gender.

Finally, I will discuss Bilqis, the Queen of Sheba. Despite the fact that she ruled over a nation, most Muslims hold leadership as improper for a woman. The Qur'an uses no terms that imply that the position of ruler is inappropriate for a woman. On the contrary, the Qur'anic story of Bilqis celebrates both her political and religious practices.

Although the verse does point out (perhaps as peculiar) that she was 'a woman' ruling (27:23), this is nothing more than a statement quoted from one who had observed her. Beyond this identification of her as a woman, no distinction, restriction, addition, limitation, or specification of her as a woman who leads is ever mentioned.

The Qur'anic story continues: she was powerful, well provided for, and 'has a magnificent throne.' (27:23). However, she and 'her people [worshipped] the sun instead of Allah' (27:24).

Solomon sends a letter to the Queen of Sheba 'In the name of Allah' and invites her and her people to submit to worship of only Allah (27:29-31). When she reads the letter, she says to her advisers: 'Lo! there has been thrown to me—a noble (*karim*) letter.' (27:29). She then asks them to advise her in this affair of hers for 'I decide no case until you [advise me (on it)]' (27:32).

Although she operates within the normal protocol and asks her advisers for consultation on this matter, she has already given an indication of her perspective by describing the letter as *karim*. Thus, her postponement of the decision on this case is not for lack of decisive ability, but for protocol and diplomacy.

'They said: we are lords of might and lords of great prowess, but it is for you to command; so consider what you will command. She said: 'Lo! kings, when they enter a township, ruin it and make the honour of its people shame. Thus will they do. But lo! I am going to send a present to them, and see with what (answer) my messengers return.' (27:33-5).

Solomon rejects the gift, stating that he has no need because Allah has given him a good position in both worldly and spiritual terms (27:36-7).

In verse 42, the story resumes after she has decided to pay him a personal visit. As she is a ruler, such a decision carries importance. It means that she has determined that there is something special and unusual about this particular circumstance which warrants her personal attention and not just that of ambassadors. Perhaps it is his first letter which is written 'in the name of Allah'[34] or because he rejects her material gift.[35]

While she is *en route* to him, Solomon has her throne (the same one mentioned above) miraculously transported to him and uses it to test her wisdom. She is led to it and asked, 'Is your throne like this?' She answers, '(It is) as though it were the very one.' (27:42). The existence of a throne so like hers, or of her throne itself, indicates something of the beyond-worldly power which Solomon commands. She goes on to say, 'Knowledge was bestowed on us in advance of this, and we have submitted to Allah (in Islam).' (27:42).[36]

Some interpret her decision to send a gift rather than to show brute strength as 'feminine' politics. I place both her worldly knowledge of peaceful politics and her spiritual knowledge of the unique message of Solomon together on the same footing to indicate her independent ability to govern wisely and to be governed wisely in spiritual matters. Thus, I connect her independent political decision—despite the norms of the existing (male) rulers—with her independent acceptance of the true faith (Islam), despite the norms of her people.

In both instances, the Qur'an shows that her judgement was better than the norm, and that she independently demonstrated that better judgement. If her politics were feminine, then her faith was feminine, which, by implication would indicate that masculinity is a disadvantage. Her faith and her politics may be specific to females, but they both were better. They indicate one who has knowledge, acts on it, and can therefore accept the truth. This demonstration of pure wisdom exhibited in the Qur'an by a woman can hopefully be exhibited by a man as well.

---

[1]It is this use of female characters which will be the focus of this chapter, because such characters demonstrate something of significance to fulfilling the Qur'anic purpose of guidance.

[2]Rahman, *Islam and Modernity*, p. 3.

[3]For a detailed discussion of this, see Abdullahi al-Na'im, *Towards an Islamic Reformation* (New York: Syracuse University Press, 1990), especially pp. 34-68.

[4]Rahman, *Islam and Modernity*, p. 5.

[5]Ibid., p. 6.

[6]Ibid., p. 7.

[7]Please see Appendix for a complete list.

[8]I will note some specific examples of this in Chapter 4, and in the Conclusion.

[9]It is still used in many Arabic-speaking countries today.

[10]Qutb, Vol. VI, p. 3621.

[11]Idem., Vol. V, p. 2817.

[12]This will be discussed in more detail in Chapter 3 concerning the meaning of the word *zawj* in verses about the Hereafter.

[13]For example, the wives of Muhammad are not permitted to remarry after his death.

[14]Individual recompense and responsibility are dealt with extensively in Chapter 3. It is sufficient here to say only that the Qur'an emphasizes that the individual will be recompensed in accordance with the deeds he or she is responsible for performing.

[15]Carol Tarvis and Carole Wade, *The Longest War: Sex Differences in Perspective*, 2nd edn. (Orlando: Harcourt Brace Jovanovich, 1984), p. 2.

[16]Qutb, Vol. II, p. 643.

[17]Idem, Vol. I, p. 523.

[18]Abbas Mahmud al-'Aqqad, *al-Mar'ah fi al-Qur'an* (Cairo: Dar al-Halal, 1962), p. 7.

[19]Which I will discuss at length in the next chapter.

[20]'See how we prefer [some] above [others], and verily Hereafter will be greater in degrees (*darajat*) and greater in preference' (17:21).

[21]For example, Pickthall translates it as 'best conduct' and 'warding off evil'; Yusuf Ali has translated it 'most righteous' and Izutsu, *God and Man*, p. 44, defines it as 'fear of God'; Maududi, Vol. XIII, p. 117, has interpreted it as 'moral excellence'. Fazlur Rahman holds strongly to his individual interpretation of this term, as opposed to others: A 'unique balance of integrative moral action ... to be squarely anchored within the moral tensions, the "limits of God" and not to "transgress" or violate the balance of those tensions or limits.' (*Major Themes*, pp. 28-9).

[22]This passage has also been interpreted to mean that each human is a descendant of a single male and a single female. See al-Zamakhshari, Vol III, p. 565. Also Maududi, vol. XIII, pp. 83 and 116.

Because of the *min* in this passage, it has also been interpreted, as I have used it, to mean that each of us is of one type or the other: male or female. See, for example, the translations of Pickthall and Yusuf Ali.

[23]It is possible to define these groups in other terms and the Arabic lends itself to this, but for the sake of simplicity, I will keep those offered by Pickthall and only distinguish them as smaller and larger groups.

[24]Or, as Sayyid Qutb, Vol. VI, p. 3344, says, 'the apparent or obvious [*zahir*] value' which men and women see is not 'the real [*haqiqi*] value', which might be hidden. Allah knows and judges by that real value.

[25]Vol. VI, p. 3348.

[26]Vol. III, p. 565.

[27]Vol. XIII, p. 117.

[28]Quoted from Muslim Ibn Majah, and translated in Vol. XIII, p. 117.

[29]This verse and its implications are discussed in detail in Chapter 3.

[30]This term means that an individual received divine communication from Allah. It is distinguished from *risalah* which indicates that an individual has received divine communication from Allah and is responsible for transmitting the information to humankind at large. The significance of this to gender is discussed in Chapter 3.

[31]For example, Ali Imran, Bani Adam.

[32]It is worth mentioning in this regard that the attitudes which promote unbearable suffering in this world for some delayed comfort in the Hereafter is contrasted here with the mercy and comfort extended to these women even above the concern for fulfilment of the divine decree.

[33]The implications of this term are discussed at length in Chapter 4. For our present discussion, it can be translated as 'co-operative and subservient'.

[34]Qutb, Vol. V, p. 2640.

[35]Maududi, Vol. 9, p. 26.

[36]I have chosen the translation by Yusuf Ali on this section of this verse over that of Pickthall's.

CHAPTER 3

The Equity of Recompense: The Hereafter in the Qur'an

*T*HE Qur'an uses the most vivid descriptions, graphic depictions and dynamic language when discussing the Hereafter. These are obviously meant to have a profound effect—during the time of revelation and for readers thereafter. The ability to sustain such an effect speaks of the pure power of language. My discussion centres on one question regarding this beautiful and impressive verbal display: Is there an essential distinction between the women and men in the Qur'anic portrayal of the Hereafter?

My attempt to answer this question will begin with a background look at this topic in the Qur'an. A further look at some particulars with regard to gender will reveal the effects of interpretation on the answer to this question.

The Hereafter and Creation

The Hereafter and Creation are two significant Unseen phenomena discussed in this book. However, unlike Creation, which is discussed with few details in the Qur'an, the Hereafter is discussed at length.

The empirical language of human communication is used for discussions of Unseen phenomena, with special rhetorical devices to reiterate that it is *other* than of this world.[1] Language is used to help project the imagination into a realm of existence about which no one has firsthand knowledge or experience. However, some of the elaborate descriptions of Paradise with regard to gender will need considerable attention.

Values and the Hereafter

The Qur'an does compare things—including people[2]—and on the basis of various criteria, it makes distinctions, determines classifications, and assigns values. In our current discussion, for example, the Hereafter is more valuable than life on earth:[3] 'The Hereafter is better (khayr) and longer lasting (abqa)' (87:17); 'The Hereafter is better (khayr) for you than the first' (93:4).[4]

The Hereafter is better because: **1**. That which lasts is better than that which passes away. While the earth is seen as limited in time, or transient—'The life of this world is but brief comfort' (13:26)—the Hereafter has no limitations in time, i.e is endless.[5] So, 'things which endure are better' (18:46). **2**. *True and pure* good is greater than apparent and/or adulterated good. On earth, pleasure and pain are mixed. The experiences of the Hereafter are all pure, unmixed. **3**. On earth things can be deceptive, giving the appearance of good when in fact they are not. The Hereafter represents *the reality*.[6] So, although the Qur'an recognizes earthly values, from Allah's perspective, the value of the Hereafter is greater than this world.[7]

Stages of the Hereafter

The Hereafter is divided into several stages, all of which are after this life. These stages are not distinguished into comprehensible terms chronologically. Sometimes they follow an apparently ordered sequence: Death, Resurrection, Judgement, and finally, Heaven or Hell. At other times they seem to combine into a single event. As with other aspects of the Unseen (al-ghayb), from the perspective of the known world (al-shahadah), this incoherency of time is understandable.

Death

Death is the first inevitable experience for each and every created being without distinction of gender, class, nationality, or time of existence. It is the first stage of the Hereafter for each being. It marks the entrance into that realm.

Each *nafs* will taste of death. And you will be paid on the day of Resurrection only that which you have fairly earned. Whoever is removed from the Fire and is made to enter Paradise, he indeed is triumphant. The life of this world is but comfort of illusion. (3:185).[8]

The term *nafs*, significant in our discussion of creation above, is also a key term in Qur'anic discussions of the Hereafter. With regard to death, it is that essential part of each being which experiences the passage from life in this known world into the unknown realms of the Hereafter. It is central to my consideration of the Hereafter because it is this term that is used to elevate the Qur'anic discussions of recompense in the Hereafter above gender distinctions.[9]

Death also separates the creatures from the Creator. Allah never dies (25:58) but 'Allah receives (men's) *anfus* at death, and that (*nafs*) which does not die (yet) in its sleep. He keeps that (*nafs*) for which he has ordained death and dismisses the rest until an appointed term' (39:42).

Qur'anic discussions of death illustrate the significance of the Hereafter to its *Weltanschauung*. As an inevitable reality, the Hereafter is an incentive for humankind to believe, to follow the guidance, and to do good deeds. The eternal realm is more valuable than this earthly transitory realm. In order to attain the best in that eternal realm, one must believe, follow the guidance and do good deeds. Thus, death is inevitable for all humans (male or female), and distinct only on the basis of the quality of one's deeds and consciousness.

Resurrection

The Resurrection is the first part of the Hereafter unknown to and/or unacceptable by the Arab audience at the time of the revelation. It also was spoken of in convincing dynamic terms:

And they say: When we are bones and fragments, shall we be raised up a new creation? Say: Be you stones or iron or some created thing that is yet greater in your thoughts. (17:49-50).

Say: Allah gives life to you, then causes you to die, then gathers you unto the Day of Resurrection, whereof there is no doubt. But most of humankind know not. (45:26).[10]

The Qur'an is emphatic that the day of Resurrection is unlike anything heretofore experienced:

O humankind! Fear your Lord. Lo! the earthquake of the Hour (of Doom) is a tremendous thing: On that day when you behold it every nursing mother will forget her nursing and every pregnant one will be delivered of her burden, and you will see humankind as drunken, yet they will not be drunken, but the Doom of Allah will be strong (upon them). (22:1-2).

On the Day when a man flees from his brother, And his mother and his father, And his wife and his children, Every man that day will have concerns enough to make him heedless (of others). (80:34-7).

Thus, the day of Resurrection is a disruption of the order of the reality which we have known and lived. However, there is one familiar aspect: the *nafs* is the vehicle which experiences this and all other events in the Hereafter. First, at the time of this Resurrection, the *nafs* will be linked or be paired with other *nufus*:[11]

When the sun is overthrown, And when the stars fall, And when the hills are moved, And when the camels ·big with young are abandoned, And when the wild beasts are herded together, And the seas rise, (*wa idhi al-nufusu zuwwijat*) And the *nufus* are *zuwwijat*.[12] (81:1-7).

The phrase *wa idhi al-nufusu zuwwijat*—and when the *nufus* are paired—brings new significance to the terms *nafs* and *zawj*. On the basis of the most essential part of the being, the *nafs* (here pluralized: *nufus*), we will be paired (*zuwwijat*) into groups or classes. Again, the term *nafs* corresponds equally to the male and the female human essence, which is the fundamental determinant of being—not gender, race, nationality, nor even religion.[13]

Although the Resurrection is a significant part of the Qur'anic eschatology, it is only this depiction of the equity of classifications on the basis of the *nafs* that concerns our discussion on male and female. There is absolute consensus among the *mufassirun* (commentators) in terms of this equity of the Resurrection. Therefore, I will not discuss other details of the Qur'anic depictions of the Resurrection.

Judgement

Existence on earth is described in the Qur'an as a trial or test to see the value of the deeds performed by individuals:

Every *nafs* will taste of death. And you will be paid on the Day of Resurrection only that which you have fairly earned.... The life of this world is but comfort of illusion. Assuredly you will be tried in your property and in your *anfus*. (3:185, 186).

Had Allah willed He could have made you one community. But that He may try you by that which He has given you (He has made you as you are). So vie with one another in good works. Unto Allah you will all return, and He will then inform you of that wherein you differ. (5:48).

At the Resurrection all humans will be brought back to life for judgement. Judgement is made by Allah, Who knows what is secret and what is manifest:

And not an atom's weight in the earth or in the sky escapes your Lord, nor what is less than that or greater than that. (10:61).

. . . and whether you make known what is in your minds or hide it, Allah will bring you to account for it. (2:284).

The Judgement will be performed in the following manner. First, all deeds of every individual are weighed on a scale. That scale indicates the *true* measure or value of things. Evil things are without merit: weightless. The weight of each good thing will be added and given multiple value.[14] Once this measurement of the deeds has been made, the results are tabulated. If the scales are light, one has been unsuccessful. If the scales are heavy, it is an indication of success.

Then, as for those whose scales are heavy (with good works), He will live a pleasant life. But as for him whose scales are light, The Bereft and Hungry One will be his mother. Ah, what will convey to you what she is! Raging fire. (101:6-11).

The final decision concerning the ultimate goal for each individual is then made. From *nafs* to *nafs* (from individual to individual) the scales on the Day of Judgement will be the same:

The weighing on that day is the true (weighing). As for those whose scale is heavy, they are successful. (7:8).

And We set a just balance for the Day of Resurrection. So that no *nafs* is wronged in aught. Though it be the weight of a grain of a mustard seed, we bring it. (21:47).

The Equity of Recompense

Lo! this life of the world is but a passing comfort, and lo! the Hereafter, that is the enduring home. Whoever does an ill deed, he will be repaid the like thereof, (*wa man 'amila salihan min dhakarin aw untha wa huwa mu'min fa ula'ika yadkhuluna al-jannah*) while whoever does right, whether male or female, and is a believer, (all) such will enter the Garden. (40:39-40).

48

I will consider these Qur'anic verses in detail and discuss syntactically the equity of recompense,[15] with a particular focus on the Arabic neuter. Through this discussion I wish to demonstrate that in order to convey eternal principles and values (i.e. universals), the Qur'an had to overcome gender constraints in the language of Arabic speakers.[16] The significant parts of this verse are:

man/ does good /*min* (from) /*dhakarin aw untha*, (male or female), and is a *mu'min* (believer) [singular masculine form] /*ula'ika* (they: masculine plural form)[17] will enter the Garden.

The word *man* is used for the interrogative 'who?' and the English 'whoever'.[18] It is one[19] of the rare Arabic terms used with both masculine and feminine with no change of form.[20] However, in the usual androcentric analysis, the *lafz* (term and form) is taken to be masculine[21], and becomes feminine (or plural or dual) in meaning only after it becomes obvious that the word applies to something feminine (or plural or dual).[22]

Restricting the *lafz* to masculine and then requiring a female or plural subject is a common feature in the language. It overrides the significance of neuter, by making it masculine and removing the reader from the potential neuter implications of a term. To alleviate such limitations, this term can be classified as neuter used equally with masculine, feminine, or plural when specified. At the least, one should understand that the use of the masculine *lafz* **represents** the neuter and therefore equally includes the female.

That *man* is used to express neutrality in the above verse is clear from the wording 'from male *or* female', which follows it. The use of *aw* 'or', rather than *wa*, 'and', indicates the individual, because it keeps the male and the female distinct, separate: *whichever* one of the two.

They are both (or either) then described as *mu'min* (a believer). In this context, we can again take the masculine singular *form* as neutral,[23] that is, not male *mu'min*, which by analogy or extension[24] includes female *mu'minah*, but neuter *mu'min*.

Finally, the plural *ula'ika* proposes inclusiveness: not only each male and female, but also, every male and female who fits the description. The significance of this analysis is that it works in conjunction with the Qur'an's emphasis on the individual in the Hereafter.

Recompense of the Individual

In the Qur'an, every stage of the Hereafter is experienced by the individual. An important part of the creation of humankind is the link between the end of an individual's life and the responsibilities which each individual fulfils in life.[25] 'Now have you come unto Us solitary as We did create you at the first.' (6:94).

With regard to the Hereafter, this individual responsibility and experience is given more emphasis. As mentioned above, it is the individual who experiences death, the transition from the living world to the Hereafter: 'Each *nafs* will taste of death.' (3:185). Then, it is the individual (*nafs*) which is classified with others of similar type (*wa idha nufus zuwwijat*). Finally, the recompense awarded is on the basis of the individual. Whether a man or a woman, each is rewarded individually according to what he or she earns, though there is a single (i.e. ungendered) scale for judgement.

Thus with regard to recompense, the Qur'an reminds us of the following points:

1. Recompense is *acquired* not through gender, but through actions performed by the individual before death.[26]

Or do those who commit ill deeds suppose that We shall make them as those who believe and do good works, the same in life and death? Bad is their judgement! And Allah has created the heavens and the earth with truth, and each *nafs* may be repaid what it has earned. And they will not be wronged. (45:21-2).

2. Although Allah oversees the Judgement and can forgive misdeeds or increase the reward for good deeds, it is not His doing that results in reward or punishment.[27] 'Lo! Allah wrongs not humankind in aught; but humankind wrong themselves (*anfus*).' (10:44).[28]

3. No one can diminish the merits earned by another; neither can anyone increase them. No one can share in the merits achieved by another, nor in the punishment which will be given.

'And guard (yourselves) against a day when no *nafs* will in aught avail another (*nafs*), nor will compensation be accepted from it, nor intercession be of use to it, nor will they be helped. (2:48).[29]

4. Despite the common misinterpretations rendered for the term *zawj*[30] there are no compensations attained or retained on the basis of one's relation to another.

5. Perhaps because of the explicit wording in the Qur'an with regard to equitable recompense, there is an unusual consensus among the commentators with regard to the absence of male/female distinctions in the Qur'anic accounts of Judgement and recompense: 'Unto men a fortune from that which they have earned, and unto women a fortune from that which they have earned.'[31] It mentions explicitly male and/or female often enough to eliminate any doubts as to its intent. 'Lo! I suffer not the work of any worker, male or female, to be lost. You proceed one from another.' (3:195).

Commentators generally use verses like 3:195 to discuss this absence of gender preference in the Hereafter. Al–Zamakhshari says that the 'worker' in this verse is clarified by the use of 'male or female' because they have a 'partnership in what Allah has promised'[32] Sayyid Qutb[33] says that 'the work' is accepted from all: males and females, because 'each of them is the same in humanness ... and on the scales'. Maududi[34] says the verse means: 'In My sight, all of you are alike as human beings and I have the same standard of justice and judgement for all, and men should not forget that women (are of/have) the same human status as they themselves have.' Thus it is inevitable that the details of my discussion will focus on some other points of contradiction to this equitable reward system.

The Final Abode

The Qur'an acknowledges our earthly values and fears. 'We verily created humankind and We know what his *nafs* whispers to him and We are nearer to him than his jugular vein.' (50:16). It states emphatically that Allah knows what is hidden and what is secret: 'Lo! Nothing in the earth or in the heavens is hidden from Allah.' (3:5). This knowledge of our subconscious feelings is demonstrated in a number of verses in the Qur'an. In particular, our fears and desires are transmitted into depictions of the afterlife. However, the Hereafter extends far beyond our imagination.

For those who do good in this world there is a good (reward) and the home of the Hereafter will be better. (16:30).

And of them is one who says 'our Lord! Give unto us in the world that which is good and in the Hereafter that which is good, and guard us from the doom of the Fire.' (2:201).

Hell

My discussion on Hell will be very brief. The descriptions of Hell depict various forms of chastisement, misery and despair, and cries of anguish. However, there is no gender distinction supplied or assumed in terminology or interpretation. Apparently the attributes of despair and misery are not gendered![35]

Lo! The guilty are immortal in hell's torment. It is not relaxed for them, and they despair therein. We wronged them not, but they it was who did the wrong. (43:74-76).

Woe unto the repudiators on that day! (It will be said to them:) Depart unto that (doom) which you used to deny; Depart unto the shadow falling threefold. Which yet is no relief nor shelter from the flame.... This is the day wherein they speak not. (77:28-31, 35).

Paradise

The Qur'an depicts Paradise in more exquisite details than any other Unseen phenomenon. In general, the depictions of Paradise are meant to entice the readers towards the afterlife. There are some forms given to the pleasures of Paradise which are specific to the audience at the time of the revelations, the desert dwellers of seventh-century Arabia. Thus the appeal of the descriptions of 'gardens with rivers flowing beneath' is greater for someone living in an arid desert environment than, perhaps, for someone living in the tropics of Malaysia.

As such, I draw an explicit relationship between the context of the revelation and some of the specific descriptions rendered. However, to confine the Qur'an to only that context would be incorrect. The 'Qur'an is from God', and 'not confined to, or exhausted by, (one) society and its history....'[36] Thus, although the perspectives of the seventh-century desert people were given significant consideration in the Qur'an's modes of expression, its eternal message is not limited to any single form of articulation. The values indicated by the text transcend the particular modes of expression. Readers from varying contexts must determine how those particulars are significant and express them in terms relevant to their own lives. Each new generation of Qur'anic readers must re-evaluate Qur'anic values and, more specifically, must determine what the expressions of Paradise mean to them.[37]

Without this double movement, we might limit the sensual descriptions of Paradise in the Qur'an to their narrowest literal meaning, rather than understand them as metaphorical indications of pleasure. We are not meant to be lost in the mere words that the Qur'an uses to depict an ineffable realm of pleasure.

Since Paradise and its pleasures are beyond human comprehension, the resemblance in these descriptions to pleasures experienced in this world must be taken analogously.[38] First, the Qur'an acknowledges the good (*khayr*)[39] in some earthly things, like wealth, power, food, family status, offspring, and women.

Beautified for mankind is love of the joys (that come) from women and offspring, and stored-up heaps of gold and silver, and horses branded (with their mark), and cattle and land. That is comfort of the life of the world. Allah! With Him is a more excellent abode. Say shall I inform you of something better than that? For those who keep from evil, with their Lord are gardens.... (3:14-15).

However, the Qur'an presents the message that these good things must be viewed with eternity in mind. They are less valuable than deemed by earthly inhabitants, especially in comparison to the eternal realm which humans cannot see.

Lo! humankind is an ingrate unto his Lord ... and lo! in the love of good (of this world) he is violent. (100:6,8).

The comfort of this world is scant; the Hereafter will be better for him who wards off evil); and you will not be wronged the down upon a date-stone. (4:77).

Note how the terms are relative to the subconscious of a particular audience (one which knows of the date-palm), yet, the notions are intended for the larger reading audience. The relationship is made between the things we enjoy on earth, in an Islamic social framework, like companionship, wealth, comfort, and power, and the greater pleasures in Paradise, to which we should direct our lives.

Companions in the Hereafter

It is clearly stated at various places in the Qur'an that one of the pleasures of Paradise will be some sort of companion. Discussions of these verses have gone to great lengths to specify the nature and number of these companions from the male perspective.[40] A clear review of the Qur'anic material on this subject requires, first and foremost, a look at chronology. During the Makkan period, the first thirteen years of revelation, the Qur'an spoke primarily to an audience of prominent patriarchs in a patriarchal society. It took into consideration their perspectives when it attempted to persuade.

The mechanisms of communication (the terms and images) employed by the Qur'an reflect that audience. They had to be convinced to change their way of thinking and their manner of living. Specifically, the Qur'an attempted: **1.** to convince them of the authenticity of the message; **2.** to demonstrate its relevance and significance; **3.** to indicate the shortcomings and weaknesses of the existing status quo; and, **4.** *to persuade or entice them through offers and threats that appealed to their nature, understanding, and experience.*

With regard to the Hereafter in particular, the Qur'an attempted first to convince them that it is real, and then to persuade them to strive for it by appealing to their nature and experiences. It is undeniable that the individuals in power had a particular attitude with regard to the importance of wealth and women (100:6-8, 3:14-15).

Descriptions of the companions in Paradise for the believers are presented on three levels. At the first level is the *hur-ul-'ayn*, which reflects the level of thinking of the Makkan community described above. The second level, which is clearly depicted during the Madinan period (and in no way similar to the Makkan period), represents the practical model of Islamic community life. Here the Qur'an uses the term *zawj*. Finally, at the third level, the Qur'an transcends both of these and expresses a perspective of companionship much greater than even these.

Hur-al-'Ayn in Qur'anic Discussions of Paradise

It is clear that the term *huri*, used in the (*idhafah*) construction *hur-al-'ayn*, meant something specific to the Jahili Arab. She was 'so called by the Arabs of the desert because of her whiteness or fairness or cleanness'. She was a woman of 'clear complexion and skin'.[41] The descriptions given of the *huri* are specific and sensual—youthful, virgin females with large dark eyes, white skin, and a pliant character—'while nowhere ... are found similar descriptions detailing, if not the beauty, at least the modest or even perhaps hidden assets of earthly wives'.[42]

The specific depiction here of the companions of Paradise demonstrates the Qur'an's familiarity with the dreams and desires of those Arabs. The Qur'an offers the *huri* as an incentive to aspire after truth. It is impossible to believe that the Qur'an intends white women with large eyes to represent a single universal description of beauty for all humankind. If we take these mythological depictions universally as the ideal female, a number of culturally specific limitations are forced on the divergent audiences of the Qur'an. The value of these particulars is extremely limited.

The Qur'an itself demonstrates the limitation of this particular depiction when the community of believers in Islam had increased in number and established itself at Madinah. After the Makkan period, the Qur'an *never* uses this term again to depict the companions in Paradise. After Madinah, it describes the companions of Paradise in generic terms: 'For those who keep from evil, with their Lord are Gardens underneath which rivers flow, and pure *azwaj*, and contentment from Allah.' (3:15).[43] Keeping in mind my arguments about the generic use of certain terms, 'believers' here are either male or female, especially since *azwaj* is used for both in the Qur'an.

Zawj in the Hereafter

Although the Qur'an is emphatic that recompense is based on the individual, some commentators disregard this when attempting to explain the term *zawj*. The use of *zawj* has been interpreted in such a manner that a 'man has the power to directly determine the fate of his spouse'.[44] Through these misinterpretations, the female *zawj* is either restricted by the limitations of her husband, or given increased rewards on the basis of his merits.

This contradicts the basic understanding in Islam, that we retain our individual responsibility for belief and disbelief when we marry,[45] and that the value of good and evil is reflected ultimately on to the individual *nafs*.[46] What is more, such interpretations contradict the Qur'anic stand against individual intercession at the Judgement.

The Qur'anic use of 'you and your *azwaj*' with regard to the Hereafter needs a closer look. First, the separation between good and evil takes precedence and the individual is recompensed only in accordance with his or her deeds:

This is the day of separation, which you used to deny. (And it is said unto the angels:) Assemble those who did wrong, together with their *azwaj* and what they used to worship. (37:21-2).

This day no soul is wronged in aught, nor are you requited aught save what you used to do. Lo! those who merit Paradise this day are happily employed, They and their *azwaj* in pleasant shade. (36: 54-6).

You who have believed Our revelations and were self-surrendered, Enter the Garden, You and your *azwaj*, to be glad. (43:69-70).

Second, the Qur'an reminds us that only those who have done right will attain rewards in Paradise, even if on earth they were related:

Our Lord! And make them enter the Gardens of Eden which You have promised them, with such of their fathers and their *azwaj* and their descendants *as do right*. (40:8).

Gardens of Eden which they enter, along with all *who do right* from their fathers, their *azwaj* and their seed. (13:23).

Thus the use of 'you and your *zawj*' means 'you and whoever is *paired* with you because of like nature, deeds, faith, etc.'

Finally, during the Madinan period[47] the use of *zawj* and *azwaj* for the companions who await believers in Paradise reflects the essential pair as discussed in Chapter 2.[48]

And give glad tidings unto those who believe and do good works, that theirs are Gardens underneath which are rivers.... There for them are pure *azwaj*. (2:25).

For those who keep from evil, with their Lord are gardens underneath which rivers flow, and pure *azwaj*, and contentment from Allah. (3:15).

And for those who believe and do good works, We shall make them enter Gardens

56

underneath which rivers flow to dwell therein for ever; there for them are pure *azwaj*—and we make them enter plentiful shade. (4:57).

The emphasis then is on partnership, friendship, comfort, and harmony in Paradise, as opposed to the isolation, loneliness, and despair of Hell. Perhaps one might be reunited with his or her earthly mate in Paradise, provided that the basis for the reunion is shared belief and good deeds, as indicated above.

It is important to clarify that most authors assume every use of the word *zawj* is equal to, or the same as, the *huri*, especially in the light of the verse that uses *huri* and the verb *zawwaja*: 'thus, we will pair them with the *hur-ul 'ayn.*' (44:54). That the term *zawwaja* means 'to join together, or, to pair up' does not equate *zawj* with *huri*, but expresses, during the Makkan period, that a man will be joined by a delightful companion according to his ideal.

What is more, some commentators use the Qur'anic statements that there will be pure *azwaj* (i.e. plural) as an indication that a pious man will go to Paradise and have multiple *huri* for his pleasure. Indeed, it is a contradiction of terms that a pious man who practises self-constraint should have multiple erotic pleasures as his objective. The absurdity of this is twofold. First, the use of the plural *azwaj* corresponds to the use of the plural preceding it: for 'believers' (and such terms). The usage is meant to indicate that companionship awaits those who believe (male and female) in their attainment of Paradise—not that each man will get multiple wives!

Second, and more significant from my analysis, each use of *zawj* and *azwaj* is not equated with *huri*, because equating the terms reduces the Qur'anic depiction of the highest reality to a single ethnocentric world-view. Such a narrow perspective cannot be criticized enough.

The Hereafter from Allah's Perspective ('inda Allah)

Finally, Paradise offers a standard at an even higher level: the perspective of Allah. From this perspective, the greatest importance of Paradise is attaining peace, ending all want, transcending all earthly limitations, and, finally, coming into the company of Allah. These highest pleasures are the same for female inhabitants of Paradise as for male. With regard to the eternal, both woman and man are equal in their potential to experience this highest transcendence. When the Qur'an offers fulfilment of desire in Paradise *'inda Allah* (from Allah's perspective), what is most important will be closeness to Allah.

So Paradise is described to the Muslim unconscious: whatever pleasures you like, such will await you in Paradise—if you restrain yourself from over-indulgence, misuse, and abuse here on earth. For the Jahili Arab patriarch, the primary audience of the Makkan period, it might be young virgin women with white skin and large, dark eyes. However, the Qur'an's descriptions of the companions of Paradise must be viewed on the basis of its entire system of justice and its objective of universal guidance, and from its own descriptions of Allah's perspective. If this is the eye through which we view the text, we will see a great deal more in it.

Summary

In summary, then, the depictions of the Hereafter which the Qur'an presents are directed to the individual. At times this individual is referred to by the term *nafs* and at other times as 'one of You'. This individual is not distinguished on the basis of gender, but on the basis of faith and deeds, which is the standard for distinctions consistently applied in the Qur'an.

Recompense is distributed with complete equity between individuals with no regard to gender. The potential to attain the best reward or to receive the ultimate punishment lies equally before women as before men. The Qur'an is emphatic and explicit about this.

Although the detailed and graphic depictions of the Hereafter—Paradise and Hell—are sometimes quite explicit, it is obvious that these descriptions are not to be taken entirely literally to the exclusion of other depictions of pleasure and pain. Rather, these graphic descriptions are the Qur'an's way of making the ineffable effable, of making the Unseen phenomena conceivable. Each will be given things similar to what they have enjoyed here.[49]

The tendency to overlook the differences between the Makkan and Madinan descriptions of Paradise has led to erroneous conclusions. The message of the Qur'an was not complete in Makkah. At Makkah, some indications were given that further development was necessary before the completion of the message in Madinah. That completed message taken in its entirety is clearer on its ultimate goals. Thus, the universal *azwaj* as companions described in the Madinan verses indicates a point in the progression, and completes the essential pairing by which humankind was created. The final goal is still higher.

The greatest value of Paradise from Allah's perspective is closeness to Him, the One. Surely the attainment of things similar to earthly pleasures represents only a lower stage in the values to be developed ultimately. This stage must be kept in its place and not extended as representative of the entire intent of the message. The more specific a description is, the more its intention is restricted.

Ultimately in the Qur'an, the reward for good is good,[50] and good and evil are not equal. From the perspective of Allah, good is infinitely more valuable than evil. It is possible that the optimism which the Qur'an seems to express with regard to the fate of humankind is directly connected to this value system.

---

[1]See Introduction.

[2]See Chapter 4 for a detailed discussion of comparisons between people.

[3]Other verses which compare these two are: 25:24, 7:169, 16:30, 17:21, and 87:17.

[4]al-Zamakhshari, Vol. 3, p. 162, says that on this earth one cannot imagine how much better the Hereafter is.

[5]Ibid., al-Zamakhshari, points out that the deeds which are to be recompensed pass away while the compensation lasts.

[6]'If you could see them when they are set before their Lord! He will say: Is not this real? They will say: Verily, by our Lord!' (6:30).

[7]I will discuss this in detail under The Hereafter *'inda Allah*.

[8]Also verses 2:180, 3:168, 4:18, 5:106, 6:61, 23:99, and 29:57, which all use some form of *nafs*, or 'one of You (pl.)' in the discussion of the inevitability of death. This will be discussed at length under Individual Recompense.

[9]See Judgement and Recompense.

[10]Also verses 32:10 and 37:58-9.

[11]This term, *nufus*, is another plural of *nafs*. Lane discusses the distinction between this plural and *anfus*, and includes in the former human and non-human breathing creatures, and in the latter only human.

[12]Here *zuwwijat* is a passive verbal form of *zawj*, i.e. to be paired or grouped or joined.

[13]That is, Muslims will not necessarily be paired with Muslims, nor Christians with Christians.

[14]Verses: 57:18, 92:23, 24:38, 35:30, 42:26

[15]Also verse (3:195)—'Lo! I suffer not the work of any worker (masculine singular), male or female, to be lost'—which is commented on by the commentators (Sayyid Qutb, *Fi Zilal al-Qur'an*, 6 vols. (Cairo: Dar al-Shuruq, 1980), Vol. I, p. 584; al-Zamakhshari, op. cit., Vol. I, p. 485; S. Abul A'la Maududi, *The Meaning of the Qur'an*, edited by A.A. Kamal and translated by Ch. Muhammad Akbar, 6th edn., 13 vols. (Lahore, Pakistan: Islamic Publication Ltd., 1983), Vol. II, 79), as indicating essentially the same point: both males and females share in the rewards of the Hereafter. However, it does not carry as many of the significant yet subtle dimensions I will comment on using this verse.

[16]See Introduction concerning the shortcomings of human language and the distinction between the language of the Arabs and Divine language.

[17]See Introduction for a detailed discussion of the neutral significata of the masculine plural form in Arabic and the Qur'an.

[18]It is also used in the conditional, but 'whoever' is still a sufficient translation.

[19]Likewise, terms like *kull* (all) and *ba'd* (some), see Dilworth B. Parkinson, 'Agreement of *ba'd* and *Kull* in Modern Literary Arabic', *Al-'Arabiyya, Journal of the American Association of Teachers of Arabic*, Vol. 8 (Spring and Autumn, 1975): pp. 52–68.

[20]See 'Abbas Hasan, *al-Nahw al-Wafi*, 4 vols. 4th edn. (Egypt: Dar al-Ma'arif, n.d.), Vol. I, p. 349: *man* is 'one of those words with a masculine singular form which can be used with or connected to and therefore conjugated by all singular, dual, plural, masculine and feminine'.

[21]See *al-Nahw al-Wafi*, Vol. I, p. 348.

[22]For example: *man hiya* means 'who is she?' and *man huwa* means 'who is he?'

[23]This is similar to the use of the masculine singular in English to denote the generic. The recent insistence on 'he/she' lacks style. Equally awkward due to unfamiliarity is the use of 'she' to denote the generic.

[24]By extension, here is the usual grammatical analysis that 'and he is a believer' implies the extension onto 'and she is a believer' (*huwa mu'min wa hiya mu'minah*). I prefer the inclusiveness of male and female in a neuter form.

[25]See Chapter 1 for a detailed discussion of individual responsibility.

[26]There are numerous verses with the same general theme, among them 3:25, 3:161, 4:110–12, 10:30, 14:51, 16:111.

[27]Qutb, op. cit., Vol. III, p. 1779, claims that neither reward nor punishment is increased; rather, one is requited for what deeds one has performed. Multiplicity in terms of good is not an *increase*; it is commensurate with the deed.

[28]Also verses 2:57, 3:117, 7:160, 9:70.

[29]Also verses, 2:48, 82:19, and 6:164, with similar wording or similar concept.

[30]A detailed discussion of the Qur'an's use of the term *zawj* with regard to the Hereafter will follow.

[31]This verse has been discussed in its entirety and within the context of verses which surround it in Chapter 2. It is not restricted to compensation in the Hereafter; however, it does include the Hereafter. The significant point here is recompense for what has been earned.

[32]Vol. I, pp. 489-90.

[33]Vol. I, p. 548.

[34]Vol. II, p. 77.

[35]Apparently, the use of the masculine plural for the depictions of Hell is never presumed to apply to them exclusively!

[36]Wan Mohd Nor Wan Daud, *The Concept of Knowledge in Islam and Its Implications for Education in a Developing Country* (London: Mansell Publisher Limited, 1989), p. 7.

[37]This method is discussed in greater detail in the Introduction taken from F. Rahman, *Islam and Modernity: Transformation of an Intellectual Tradition* (Chicago: The University of Chicago Press, 1981).

[38]See 'The Special Case of Women and Children in the Afterlife', p. 116.

[39]For example, verses 38:32, 68:12.

[40]For a delightful critique of these, see Fatima Mernissi's 'Women in Moslem Paradise' (New Delhi: Kali for Women, 1986).

[41]Edward William Lane, *An Arabic—English Lexicon*, 8 vols. (London: Librarie du Liban, 1980), Part *Jim To Ta*, p. 666.

[42]Fatna A. Sabbah, *Woman in the Muslim Unconscious*, translated by Mary Jo Lakeland from the French *La Femme dans l'inconscient Musulman* (New York: Pergamon Press, 1984), p. 95.

[43]See also 'The Special Place for Women and Children in the Afterlife', p. 165.

[44]Jane I. Smith and Yvonne Y. Haddad, 'Women in the Afterlife: Islamic View as Seen from the Qur'an and Traditions', in *Journal of the American Academy of Religion*, Vol. 43 (1975): p. 45.

[45]See detailed discussion of *surat-al Tahrim* above.

[46]See, for example, verses 6:51, 6:70, and 32:4.

[47]As opposed to the use of the term *hur-al-'ayn* in the Makkan period of revelation, see discussion below.

[48]See Chapter 1.

[49]See verses 2:24 and 2:26.

[50]See verse 44:60.

CHAPTER 4

Rights and Roles of Woman:
Some Controversies

IT would be impossible to have a discussion on any topic in the
Qur'an which would exhaust the material covered in the text
itself. Nor would it be possible to conclude in definitive terms the
significance of all the material in the Qur'an concerning humankind on earth. The text was revealed to the inhabitants of the earth,
while they inhabited the earth, and we are all on the earth as we
read and discuss the text. As such, our earthly existence transforms
our perceptions of the text and is equally potentially transformed by
the text. More importantly, because of the Qur'an's intention to
guide the affairs of humans, a certain emphasis is placed on understanding and applying the text while we are here on earth.

In my consideration of woman on earth from the Qur'anic perspective, there are certain problems inherent in our understanding
of what the Qur'an depicts. Our operations on the earth are shaped
by our world-view (and vice versa). We have not yet attained the
Qur'anic utopia. Whenever Qur'anic support is given for conflict-
ing opinions on how to operate in this world, controversies arise.
Many popular or dominant ideas about the role of woman do not
have sanction from the Qur'an. Pointing these out causes problems,
not so much with logical analysis of the text, but with application of
the new analysis in the context in which Muslim societies operate.

Hermeneutics of any text must confront three different aspects
in order to support its conclusions: **1**. the context in which the text
was written (in the case of the Qur'an, in which it was revealed);
2. the grammatical composition of the text (how it says what it
says); and **3**. the whole text, its *Weltanschauung* or world-view.

Often, differences of opinion can be traced to variations in emphasis between these three aspects.

I will discuss selected concepts, terms or verses from these perspectives: **1.** There is no inherent value placed on man or woman. In fact, there is no arbitrary, pre-ordained and eternal system of hierarchy. **2.** The Qur'an does not strictly delineate the roles of woman and the roles of man to such an extent as to propose only a single possibility for each gender (that is, women must fulfil this role, and *only* this one, while men must fulfil that role and only men *can* fulfil it).

To demonstrate these points, I will make a detailed analysis of Qur'anic passages which have been interpreted to imply the superiority of males over females. In doing this, I will demonstrate a more integrated communal perspective on the rights and responsibilities of the individual in society using certain Qur'anic concepts.

Overall, my analysis tends to restrict the meaning of many passages to a particular subject, event, or context. These restrictions are based on the context of the verses or on application of general Qur'anic concepts of justice towards humankind, human dignity, equal rights before the law and before Allah, mutual responsibility, and equitable relations between humans.

One other aspect of my consideration will be to demonstrate the significance of chronological developments in the Qur'an. The Qur'an sets forth a logical progression with regard to the development of human interactions, morality, and ethics,[1] as reflected by the growth and development of the community of Muslims who lived concurrent with the revelation. The significance of this with regard to application of the precepts concerning the role of woman will be demonstrated.

Functional Distinctions on Earth

In Chapter 3, I demonstrated how the Qur'an treats woman as an individual in the same manner as it treats man as an individual. Their only distinction is on the basis of *taqwa* (God-conscious piety). *Taqwa* is not determined by gender. The Qur'an also focuses on how we function in society. It acknowledges that we operate in social systems with certain functional distinctions. The relationship that the Qur'an shows between these worldly distinctions and *taqwa* is important in my consideration of equity among people. More importantly, functional distinctions included in the Qur'an have been used to support the idea of inherent superiority of men over women.

Functional distinctions are indicators of roles and role expectations. To what extent does the Qur'an delineate functions for each gender? Are there certain exceptions and exclusions for males or females? Does the Qur'an value certain functions above others?

Woman is not just Biology

Because woman's primary distinction is on the basis of her child-bearing ability, it is seen as her primary function. The use of 'primary' has had negative connotations in that it has been held to imply that women can only be mothers. Therefore, women's entire up-bringing must be to cultivate devoted wives and ideal mothers in preparation for this function.

There is no term in the Qur'an which indicates that child-bearing is 'primary' to a woman. No indication is given that mothering is her exclusive role. It demonstrates the fact a woman (though certainly not all women) is the exclusive human capable of bearing children. This capacity is essential to the continuation of human existence. This function becomes primary only with regard to the continuity of the human race. In other words, since only the woman can bear children, it is of primary importance that she does.

Although it does not restrict the female to functioning as a mother, the Qur'an is emphatic about the reverence, sympathy, and responsibility due to the female procreator. 'O humankind . . . have *taqwa* towards Allah in Whom you claim your rights of one an-other, and (have *taqwa*) towards the wombs (that bore you).' (4:1). This verse is often interpreted as indicating respect for women in general.[2] I specify this verse as indicating respect for the needed procreative capacity of women. I do not diminish respect from wo-men as a class, but I do specify, from the Qur'anic perspective, the significance of the function of child-bearing, which is exclusively performed by women.[3] The reverence given to the fulfilment of this function helps to explain how the Qur'an explicitly delineates a function for males which creates a balance in human relations.[4]

No other function is similarly exclusive to one gender or the other. This brings to mind the popular misconception that since only males have had the responsibility of *risalah*[5], it indicates something special about that class. Both men and women have been included in divine communication as the recipients of *wahy*[6], but there is no Qur'anic example of a woman with the responsibility of *risalah*. However, all those chosen for this responsibility were exceptional.

This is not a biological association with males representing their primary function and expressing a universal norm for all men. In fact, given the difficulty they have faced in getting others to accept the message when these exceptional men have come from poor classes, the likelihood of failure for the message might have been greater if women, who are given so little regard in most societies, were selected to deliver the message. It is a strategy for effectiveness, not a statement of divine preference.[7]

Besides the two functions discussed above, every other function has real or potential participation by both males and females. However, there is still a wide range of functional distinctions between individuals considered in the Qur'an. The questions that must be asked then are: What is the value of the functional distinctions between individuals? Do these functional distinctions and the values placed on them delineate specific values for males and females in society? Are these values intra-Qur'anic or extra-Qur'anic?

In particular, several verses from the Qur'an have frequently been used to support the claims of the inherent superiority of males over females. These verses contain two terms which have been used to indicate value in the functional distinctions between individuals and groups on earth. I will review these terms, how they have been used in the Qur'an, and in the overall context of Qur'anic justice.

The first term is *darajah* (pl. *darajat*), 'step, degree or level'. A *darajah* exists not only here on earth between people but also between the Hereafter and earth,[8] between levels in Heaven and in Hell.

The other term, *faddala* is often used in conjunction with *darajat*. I have translated *faddala* 'to prefer', with a verbal noun (*tafdil*) meaning 'preference'. Often the preference given is spoken of in terms of *fadl*, which I translate as (Allah's) 'benevolence'.

Darajah

An individual or group can earn or be granted a *darajah* over another. The Qur'an specifies, for example, that by striving in the way of Allah with one's wealth and one's person (4:95) or by immigrating for Allah (9:20), one can obtain a *darajah*. However, most often the *darajah* is obtained through an unspecified category of doing 'good' deeds (20:75, 6:132, 46:19).

Distinguishing between individuals or groups on the basis of 'deeds' involves problems with regard to the value of women in society and as individuals. Although the Qur'an distinguishes on the basis of deeds, it does not set values for particular deeds. This leaves each social system to determine the value of different kinds of deeds at will. They have always done this and 'every society has distinguished men's work from women's work'.[9] The problem is that 'Men's work is usually regarded as more valuable than women's work, no matter how arbitrary the division of labor'.[10]

On the one hand, the Qur'an supports distinctions on the basis of deeds, but on the other hand, it does not determine the actual value of specific deeds. This leads to the interpretation that the Qur'an supports values of deeds as determined by individual societies. Actually, the Qur'an's neutrality allows for the natural variations that exist.

With regard to the *darajah* obtained through deeds, however, the Qur'an has stipulated several points which should affect evaluation in society. First, all deeds performed with *taqwa* are more valuable. Second, 'Unto men a fortune from that which they have earned, and unto women a fortune from that which they have earned.' (4:32). The deeds may be different, but recompense is given based on what one does. It does not matter how the deeds are divided between the males and the females in a particular social context.

Another implication of a 'fortune from what one earns' is that whenever anyone performs tasks normally attributed to the other gender in addition to his or her own normal tasks, he or she will earn an additional reward. For example, Moses meets two women from Madyan, where ordinarily the males tended the animals. However, because there was no able-bodied male in the family to perform this task according to the norm (the father being an old man), the women were required to be *extraordinarily* useful.

There is no indication that these women were immoral in their performance of this task, because fulfiling the tasks needed for survival takes precedence over socially determined roles. Similarly, in post-slavery America, the Black female was given employment instead of the Black male. In many families, she became the sole supporter. This necessity, in addition to her fulfillment of the ordinary tasks of bearing and rearing children, should have given her more. A flexible perspective on the fulfilment of necessity would have benefited her. Instead, she was subject to a double burden and, often, violence at home from a husband who felt displaced.

Each social context divides the labour between the male and the female in such a way as to allow for the optimal function of that society. The Qur'an does not divide the labour and establish a monolithic order for every social system which completely disregards the natural variations in society. On the contrary, it acknowledges the need for variations when it states that the human race is divided 'into nations and tribes that you might know one another'. (49:13). Then it gives each group, and each member of the group—the males and the females—recompense in accordance to deeds performed.

This is an important social universal in the Qur'an. It allows and encourages each individual social context to determine its functional distinctions between members, but applies a single system of equitable recompense which can be adopted in every social context. This is also one reason why certain social systems have remained stagnant in their consideration of the potential roles of women. The Qur'an does not specifically determine the roles, and the individual nations have not considered all the possibilities.

As for the *darajah* which is 'given' by Allah, it is even more illusive than the *darajah* for unspecified deeds. There is a distinction on the basis of knowledge: 'Allah will exalt those who believe among you, and those who have knowledge, to high ranks [*darajat*]' (58:11). 'We raised by grades [*darajat*] (of mercy) whom We will, and over all endued with knowledge there is one more knowing.' (12:76).

There are also social and economic distinctions: 'We have apportioned among them their livelihood in the life of the world, and raised some of them above others in ranks [*darajat*] that some of them may take labour from others; and the mercy of Allah is better than (the wealth) that they amass.' (43:32).[11] It is also clear, however, that wealth is not a 'real' distinguishing characteristic, but a functional distinction apparent to humankind and valued within society.

The *darajah* given by Allah serves another significant function— to test the inhabitants of the earth: 'He it is Who has placed you as viceroys of the earth and has exalted some of you in ranks [*darajat*] above others, that He may try you by (the test of) that which He has given you.' (6:165).

Finally, it is necessary to discuss the one verse which distinguishes a *darajah* between men and women:

Women who are divorced shall wait, keeping themselves apart, three (monthly) courses. And it is not lawful for them that they conceal that which Allah has created in their wombs if they believe in Allah and the Last Day. And their husbands would do better to take them back in that case if they desire a reconciliation. And [(the rights) due to the women are similar to (the rights) against them, (or responsibilities they owe) with regard to] the *ma'ruf*, and men have a degree [*darajah*] above them (feminine plural). Allah is Mighty, Wise. (2:228).

This verse has been taken to mean that a *darajah* exists between all men and all women, in every context. However, the context of the discussion is clearly with regard to divorce: men have an advantage over women. In the Qur'an the advantage men have is that of being individually able to pronounce divorce against their wives without arbitration or assistance. Divorce is granted to a woman, on the other hand, only after intervention of an authority (for example, a judge).

Considering the details given, *darajah* in this verse must be restricted to the subject at hand.[12] To attribute an unrestricted value to one gender over another contradicts the equity established throughout the Qur'an with regard to the individual: each *nafs* shall have in accordance to what it earns. Yet, the verse is presumed to state what men have believed and wanted others to believe: that society operates hierarchically with the male on top.

Finally, this verse states: '[(the rights) due to the women are similar to (the rights) against them, (or responsibilities they owe) with regard to] the *ma'ruf.*' The term *ma'ruf* occurs in other instances with regard to the treatment of women in society. Pickthall translates it as 'kindness', but its implications are much wider than that. It is a passive participle of the verbal root 'to know', and as such indicates something 'obvious', 'well known' or 'conventionally accepted'.[13] However, with regard to treatment, it also has dimensions of equitable, courteous and beneficial[14].

In this verse (2:228), it precedes the *darajah* statement to indicate its precedence. In other words, the basis for equitable treatment is conventionally agreed upon in society. With regard to this, the rights and the responsibilities of the woman and the man are the same. Again, the expression places a limitation rather than a universal perspective on this issue because convention is relative to time and place.

Faddala

As with *darajah*, the Qur'an states explicitly that Allah has preferred [*faddala*] some of creation over others. Like *darajah*, this preference is also discussed in specific terms. First, humankind is preferred over the rest of creation (17:70). Then, occasionally, one group of people have been preferred over another.[15] Finally, some of the prophets are preferred over others (2:253, 6:86, 17:55). It is interesting to note, however, that 'preference' is not absolute. Although the Qur'an states that some prophets are preferred over others, it also states that *no distinction* is made between them (2:285). This indicates that, in the Qur'anic usage, preference is relative.

Like *darajah, faddala* is also given to test the one to whom it is given. Unlike *darajah*, however, *faddala* cannot be earned by performing certain deeds. It can only be given by Allah, Who has it and grants it to whom He wishes and in the form He wishes. Others do not have it and cannot give it. They can only be recipients of His *fadl.*

69

With regard to *faddala*, men and women, the following verse is central:

Men are [*qawwamuna 'ala*] women, [on the basis] of what Allah has [preferred] (*faddala*) some of them over others, and [on the basis] of what they spend of their property (for the support of women). So good women are [*qanitat*], guarding in secret that which Allah has guarded. As for those from whom you fear [*nushuz*], admonish them, banish them to beds apart, and scourge them. Then, if they obey you, seek not a way against them. (4:34).

Needless to say, this verse covers a great deal more than just preference. This is classically viewed as the single most important verse with regard to the relationship between men and women: 'men are *qawwamuna 'ala* women'. Before discussing this, however, I want to point out that this correlation is determined on the basis of two things: **1**. what 'preference' has been given, and **2**. 'what they spend of their property (for support of women),' i.e. a socio-economic norm and ideal.

The translation I have inserted, 'on the basis of,' comes from the *bi*[16] used in this verse. In a sentence, it implies that the characteristics or contents before *bi* are determined 'on the basis' of what comes after *bi*. In this verse it means that men are *qawwamuna 'ala* women only if the following two conditions exist. The first condition is 'preference', and the other is that they support the women from their means. 'If either condition fails, then the man is not '*qawwam*' over that woman'.[17]

My first concern then is *faddala*. The verse says the position between men and women is based on 'what' Allah has preferred. With regard to material preference, there is only one Qur'anic reference which specifies that Allah has determined for men a portion greater than for women: inheritance.[18] The share for a male is twice that for the female (4:7) within a single family. The absolute inheritance for all men will not always be more than that for all women. The exact amount left depends on the family's wealth in the first place.

In addition, if verse 4:34 refers to a preference demonstrated in inheritance, then such a materialistic preference is also not absolute. This connection is often favoured because the other condition for *qiwamah* is that 'they spend of their property (for the support of women)'. Thus, there is a reciprocity between privileges and responsibilities. Men have the responsibility of paying out of their wealth for the support of women, and they are consequently granted a double share of inheritance.

However, it cannot be overlooked that 'Many men interpret the above passage' as an unconditional indication of the preference of men over women. They assert that 'men were created by God superior to women (in strength and reason)'.

However, this interpretation, ... is (i) unwarranted and (ii) inconsistent with other Islamic teachings ... the interpretation is unwarranted because there is no reference in the passage to male physical or intellectual superiority.[19]

Faddala cannot be unconditional because verse 4:34 does not read 'they (masculine plural) are preferred over them (feminine plural)'. It reads '*ba'd* (some) of them over *ba'd* (others)'. The use of *ba'd* relates to what obviously has been observed in the human context. All men do not excel over all women in all manners. Some men excel over some women in some manners. Likewise, some women excel over some men in some manners.[20] So, whatever Allah has preferred, it is still not absolute.

If 'what' Allah has preferred is restricted to the material (and specifically inheritance), then the extent and nature of the preference is explained by the Qur'an. Even if 'what' Allah has preferred is more than just the preference given in inheritance, it is, nevertheless, still restricted to 'some of them' over 'some others' by the wording in this context:

'men are '*qawwamun*' over women in matters where God gave *some* of the men more than *some* of the women, *and* in what the men spend of their money, then clearly men as a class are not '*qawwamun*' over women *as a class*.[21]

However, further understanding of this distinction requires further explanation of *qawwamuna 'ala*. What does it mean, and what are the parameters of its application?

As for the meaning, Pickthall translates this as 'in charge of'. Al-Zamakhshari[22] says it means that 'men are in charge of the affairs of women'. Maududi[23] says 'Men are the managers of the affairs of women because Allah has made the one superior to the other. . . .' Azizah al-Hibri objects to any translation which implies that men are protectors or maintainers because 'The basic notion here is one of moral guidance and caring'[24] and also because:

... only under extreme conditions (for example, insanity) does the Muslim woman lose her right to self-determination. . . . Yet men have used this passage to exercise absolute authority over women. They also use it to argue for the male's divinely ordained and inherent superiority.[25]

Some questions beg asking concerning the parameters of application: Are all men *qawwamuna 'ala* all women? Is it restricted to the family, such that the men of a family are *qawwumuna 'ala* the women of that family? Or, is it even more restricted, to the marital tie, such that only husbands are *qawwumuna 'ala* wives? All of these possibilities have been given.

Generally, an individual scholar[26] who considers *faddala* an unconditional preference of males over females does not restrict *qiwamah* to the family relationship but applies it to society at large. Men, the superior beings, are *qawwamuna 'ala* women, the dependent, inferior beings.

Sayyid Qutb[27], whose discussion I will consider at length, considers *qiwamah* an issue of concern for the family within society. He restricts verse 4:34, in some ways, then, to the relationship between the husband and the wife. He believes that providing for the females gives the male the privilege of being *qawwamuna 'ala* the female.

He gives *qiwamah* a decided dimension of material maintenance. The rationale behind restricting this verse to the context of husband and wife is partly due to the fact that the remainder of the verse discusses other details of concern to the marital relationship. In addition, the following verse uses the dual, indicating that it is concerned with the context between the two: the husband and wife. However, preceding verses discuss terms of relations between male members of society and female members of society.

I apply this verse to society at large—but not on the basis of inherent superiority of men over women, or of Allah's preference of men over women. Rather, I extend the functional relationship, which Sayyid Qutb proposes between the husband and the wife, towards the collective good concerning the relationship between men and women in society at large. My main consideration is the responsibility and right of women to bear children.

Sayyid Qutb says, 'The man and the woman are both from Allah's creation and Allah ... never intends to oppress anyone from

His creation.'[28] Both the man and the woman are members of the most significant institution of society, the family. The family is initiated by marriage between one man and one woman. Within the family, each member has certain responsibilities. For obvious biological reasons, a primary responsibility for the woman is childbearing.

The child-bearing responsibility is of grave importance: human existence depends upon it. This responsibility requires a great deal of physical strength, stamina, intelligence, and deep personal commitment.[29] Yet, while this responsibility is so obvious and important, what is the responsibility of the male in this family and society at large? For simple balance and justice in creation, and to avoid oppression, his responsibility must be equally significant to the continuation of the human race. The Qur'an establishes his responsibility as *qiwamah*: seeing to it that the woman is not burdened with additional responsibilities which jeopardize that primary demanding responsibility that only she can fulfil.

Ideally, *everything* she needs to fulfil her primary responsibility comfortably should be supplied in society, by the male: this means physical protection as well as material sustenance. Otherwise, 'it would be a serious oppression against the woman'.[30]

This ideal scenario establishes an equitable and mutually dependent relationship. However, it does not allow for many of today's realities. What happens in societies experiencing a population overload, such as China and India? What happens in capitalistic societies like America, where a single income is no longer sufficient to maintain a reasonably comfortable life-style? What happens when a woman is barren? Does she still deserve *qiwamah* like other women? What happens to the balance of responsibility when the man cannot provide materially, as was often the case during slavery and post-slavery US?

All of these issues cannot be resolved if we look narrowly at verse 4:34. Therefore, the Qur'an must eternally be reviewed with regard to human exchange and mutual responsibility between males and females. This verse establishes an ideal obligation for men with regard to women to create a balanced and shared society. This responsibility is neither biological nor inherent, but it is valuable. An attitude inclined towards responsibility must be cultivated. It is easy enough to see the cases in which it has not been acquired.

However, such an attitude should not be restricted to mere material *qiwamah*. In broader terms, it should apply to the spiritual, moral, intellectual, and psychological dimensions as well. Such a perspective on *qiwamah* will allow men to truly fulfil their *khilafah* (trusteeship) on the earth, as ordained by Allah upon human creation. Such an attitude will overcome the competitive and hierarchical thinking which destroys rather than nurtures.

Men are encouraged to fulfil their trusteeship of the earth—especially in relationships with women, the child-bearers and traditional caretakers. What women have learned through bearing and caring for children, men can begin to experience, starting with their attitudes to and treatment of women.

Nushuz: Disruption of Marital Harmony

Finally, with regard to this verse, I will discuss whether this portion,

So good women are *qanitat*, guarding in secret that which Allah has guarded. As for those from whom you fear [*nushuz*], admonish them, banish them to beds apart, and scourge them. Then, if they obey you, seek not a way against them

means that a woman *must* obey her husband, and if she does not, he can beat her (here translated 'scourge'). I believe the passage intends to provide a means for resolving disharmony between husband and wife.

First, the word *qanitat* used here to describe 'good' women, is too often falsely translated to mean 'obedient', and then assumed to mean 'obedient to the husband'. In the context of the whole Qur'an, this word is used with regard to both males (2:238, 3:17, 33:35) and females (4:34, 33:34, 66:5, 66:12).[31] It describes a characteristic or personality trait of believers towards Allah. They are inclined towards being co-operative with one another and subservient before Allah. This is clearly distinguished from mere obedience between created beings which the word *ta'a* indicates.

Sayyid Qutb points out that this choice of words indicates that the Qur'an intends there to be a personal emotional response rather than the external 'following of orders' which the *ta'a* (obey) would suggest.[32] As for the use of that word *ta'a* and the remainder of the verse, 'As for those (feminine plural) from whom you fear *nushuz* ...', it should first be noted that the word *nushuz* likewise is used with both males (4:128) and females (4:34), although it has been defined differently for each.[33] When applied to the wife, the term is usually defined as 'disobedience to the husband'. With the use of *ta'a* that follows. Others have said this verse indicates that the wife must obey the husband.

However, since the Qur'an uses *nushuz* for both the male and the female, it cannot mean 'disobedience to the husband'. Sayyid Qutb explains it as a state of disorder between the married couple.[34] In case of disorder, what suggestions does the Qur'an give as possible solutions? There is **1**. A verbal solution: whether between the husband and wife (as here in verse 4:34) or between the husband and wife with the help of arbiters (as in 4:35, 128). If open discussion fails, then a more drastic solution: **2**. separation is indicated. Only in extreme cases a final measure: **3**. the 'scourge' is permitted.

With regard to regaining marital harmony, the following points need to be raised. First, the Qur'an gives precedence to the state of order and emphasizes the importance of regaining it. In other words, it is not a disciplinary measure to be used for disagreement between spouses. Second, if the steps are followed in the sequential manner suggested by the Qur'an, it would seem possible to regain order before the final step. Third, even if the third solution is reached, the nature of the 'scourge' cannot be such as to create conjugal violence or a struggle between the couple because that is 'un-Islamic'.[35]

It appears that the first measure is the best solution offered and the one preferred by the Qur'an, because it is discussed in both instances of the word *nushuz*. It is also in line with the general Qur'anic principle of mutual consultation, or *shura*, being the best method for resolving matters between two parties. It is obvious that the Qur'an intends a resolution of the difficulties and a return to peace and harmony between the couple when it states: '...it is no sin for the two of them if they make terms of peace between themselves. *Peace is better.*' (4:128). It is peace and 'making amends' (4:128) that are the goals, not violence and forced obedience.

The second solution is, literally, to 'banish them to beds apart'.

First, the significance of 'beds apart' is possible only when the couple continually shares a bed (unlike polygamy when husband and one wife do not), otherwise, this would not be a meaningful measure. In addition, 'beds apart' indicates that at least one night should pass in such a state. Therefore, it is a cooling-off period which would allow both the man and the woman, separately, to reflect on the problem at hand. As such, this measure also has equally mutual implications.

As one night apart can lead to many nights apart before any resolution is made, this separation could go on indefinitely. This does not indicate that a man should then begin to physically abuse his wife. Rather, it allows for a mutually found peaceable solution, or a continued separation—divorce. Divorce also requires a waiting period, and beds apart is characteristic of that waiting. Thus, this measure can be taken as part of the overall context of irreconcilable differences between the married couple.

It cannot be overlooked, however, that verse 4:34 does state the third suggestion using the word *daraba*, 'to strike'. According to *Lisan al-'Arab* and *Lanes's Lexicon*, *daraba* does not necessarily indicate force or violence. It is used in the Qur'an, for example, in the phrase '*daraba Allah mathalan . . .*' ('Allah *gives* or *sets* as an example. . .'). It is also used when someone leaves, or 'strikes out' on a journey.

It is, however, strongly contrasted to the second form, the intensive, of this verb—*darraba*: to strike repeatedly or intensely. In the light of the excessive violence towards women indicated in the biographies of the Companions and by practices condemned in the Qur'an (like female infanticide), this verse should be taken as prohibiting unchecked violence against females. Thus, this is not permission, but a severe restriction of existing practices.

Finally, the problem of domestic violence among Muslims today is not rooted in this Qur'anic passage. A few men strike their wives after completely following the Qur'anic suggestions for regaining marital harmony. The goal of such men is harm, not harmony. As such, after the fact, they cannot refer to verse 4:34 to justify their action.

Finally, the word *ta'a* in this verse needs a contextual consideration. It says 'if they obey (*ta'a*) you do not seek a way against them.'

For the women, it is a conditional sentence, not a command. In the case of marriages of subjugation—the norm for Muslims and non-Muslims at the time of the revelation—wives were obedient to husbands. The husbands are commanded 'not to seek a way against' wives who are obedient.[36] The emphasis is on the male's treatment of the female.

The Qur'an never orders a woman to obey her husband. It never states that obedience to their husbands is a characteristic of the 'better women' (66:5), nor is it a prerequisite for women to enter the community of Islam (in the Bay'ah of the women: 60:12). However, in marriages of subjugation, wives did obey their husbands, usually because they believed that a husband who materially maintains his family, including the wife, deserves to be obeyed. Even in such cases, the norm at the time of the revelation, no correlation is made that a husband should beat his wife into obedience. Such an interpretation has no universal potential, and contradicts the essence of the Qur'an and the established practices of the Prophet. It involves a severe misreading of the Qur'an to support the lack of self-constraint in some men.

With regard to the relationship between maintenance and obedience, it can be observed that even husbands who are unable or unwilling to provide for their wives, believe they should be obeyed. In fact, this widespread characteristic of Muslim marriage is only one example of the association of men as natural leaders deserving obedience.

This belief in the need to obey the husband is a remnant of marriages of subjugation and is not exclusive to Muslim history. It has not progressed, although today couples seek partners for mutual emotional, intellectual, economic, and spiritual enhancement. Their compatibility is based on mutual respect and honour, not on the subservience of the female to the male. The family is seen as a unit of mutual support and social propriety, not an institution to enslave a woman to the man who buys her at the highest price and then sustains her material and physical needs only, with no concern for the higher aspects of human development.

If the Qur'an was only relevant to this single marriage type, it would fail to present a compatible model to the changing needs and requirements of developing civilizations worldwide. Instead, the Qur'anic text focuses on the marital norm at the time of revelation, and applies constraints on the actions of the husbands with regard to wives. In the broader context, it develops a mechanism for resolving difficulties through mutual or extended consultation and arbitration.

In conclusion, the Qur'an prefers that men and women marry (4:25). Within marriage, there should be harmony (4:128) mutually built with love and mercy (30:21). The marriage tie is considered a protection for both the male and the female: 'They (feminine plural) are raiment for you (masculine plural) and you are raiment for them.' (2:187). However, the Qur'an does not rule out the possibility of difficulty, which it suggests can be resolved. If all else fails, it also permits equitable divorce.

The Significance of Context and Chronology in Qur'anic Social Reforms for women

With regard to matters related to the subject of women, like divorce, it is apparent that the order in which verses were revealed in the Qur'an is more significant than for some other issues. The Qur'an responded to particular circumstances in Arabia at the time of the revelation. It is interesting to note that there are no references made in the Qur'an to specific rights, responsibilities and treatment of women until the Madinan period. 'They consult you concerning women. Say: Allah gives you decree concerning them....' (4:127). If a woman was mentioned in the Makkan period, she was a generic example for all humankind.

In the Madinan period of revelation, the particular social reforms introduced were related to the existing practices. In this respect, it is also interesting to note that most reforms were for the benefit of the females. No equivalent reforms benefiting the males were instituted, ostensibly because the existing androcentric norm greatly provided for male needs and wants. The Qur'anic responses were directed towards the pressing need for reforms regarding women.

Although the Qur'an responded to the need for reform with regard to women in the Madinan *surah*, these responses have implications for later communities. However, the particulars on which these verses focus provide clear indications about the norms in that context. In matters regarding gender, seventh-century Arabia was far from ideal. However, even Qur'anic reforms were not fully implemented. Such rapid changes met with some difficulty and reluctance. This observation has been made by twentieth-century authors with regard to some of the details in Qur'anic social reforms.[37] We will examine these details and some of the modern observations below.

Divorce

Divorce is a lawful option for irreconcilable differences between married couples. However, the condition discussed above, which allows the male a *darajah* or an advantage over the female, has been pointed out as indicating an inequity in the Qur'an—that is, men have the the power of repudiation. Unlike women, they may state: 'I divorce you' to begin the divorce procedure.

This power of the male has received some attention in modern Islamic legal reforms. In some instances, the male is required to come before the courts prior to such repudiation. In Malaysia, for example, both the husband and the wife who find difficulty in their marriage must come before the court to express their complaints. The court then acts as, or assigns, an arbiter (4:34, 35, and 128) for counselling.

Because this is similar to the traditional position for the woman, she faces no additional problem. However, when any stipulation is made in regard to what is believed to be a male privilege, men reject it as being too difficult; such a hardship should not be imposed over their rights, thus decreasing the empowerment they feel. Such a stipulation creates greater parity between the rights and responsibilities of both. The males are able to experience the matter from the point of view of the woman. Yet few men use this experience to provide for a more mutually beneficial solution.

Another consideration with regard to this verse again involves the evolution of marriages. Since women are no longer the subjects of marriage, but full, willing partners, our focus shifts to the broader

Qur'anic wisdom which aims at harmonious reconciliation: '... it is no sin for the two of them if they make terms of peace between *themselves. Peace is better.*' (4:128). 'Either take them back on equitable terms or set them free on equitable terms, do not take them back to injure them, or to take unfair advantage ...' (2:231).

Again, this consideration of repudiation is clear in view of the practices as they existed at the time of revelation, and not only for Muslim marriages. There is no indication that the unilateral right to repudiation needs be continued, or if continued, that it need be only for the husband. Although the Qur'an stipulates conditions for equitable separation or reconciliation, it does not make a rule that men *should* have uncontrolled power of repudiation. Men *did* have this power, over which the Qur'an places conditions and responsibilities.

Although the Qur'an makes no reference to women repudiating their husbands, it has been used to conclude that they cannot. This later conclusion was drawn in contrast to the custom in pre-Islamic times when a woman had simply to turn the entrance of her tent to face another direction to indicate her repudiation of conjugal relations with a man. There is no indication in the Qur'an that all power of repudiation must be removed from women.

What is important is mutual and peaceable reconciliation or separation. The Qur'an applies explicit measures to prevent abandonment and misuse of women, who were subject to the whims of husbands in marriage and divorce. It also stipulates measures of protection for the honour of both lines of parentage.

Patriarchy

The established order within the Arabian peninsula at the time of the revelation was patriarchal: a 'culture built on a structure of domination and subordination ...' which 'demands hierarchy'.[38] It was a culture with an androcentric bias, one where the male and the male experience are looked upon as the norm. (With such a bias, statements like 'All men are created equal', are left to individual whims which can interpret the term 'men' as human beings including women, or as exclusive to male humans.)

In androcentric cultures, females are looked upon in terms of their utility to men, which is primarily reproductive. That such a cultural bias was the context of Qur'anic revelation has serious implications for later communities which try to understand the social ideal the Qur'an was attempting to establish in that community. The Qur'an's accommodation of various social contexts has been viewed as implying support for the particular social order that existed in seventh-century Arabia.

Some argue that Islam is essentially a variation of patriarchal ideology. Others argue that Islam is above worldly ideology, including Patriarchy; for as the word of God,[39] it transcends all ideology. Among these, we can distinguish two groups: those who believe that Islam as it is today is fair and just to women, and those who believe that Islam *as it is practised today* is utterly patriarchal, but that *true* Islam is not.[40]

The implication of this patriarchal context must be understood in relation to the greater Qur'anic principles and their ultimate intent of harmonious and equitable relationships in society. It is this ultimate Qur'anic intent which reveals its true 'spirit'. Muslim thinkers in the twentieth century use this Qur'anic 'spirit' to argue against literal application of some Qur'anic statements.

It is clear through the chronological progression of the text, that the Qur'an sets out guidelines. Following these guidelines to their natural conclusions will not lead backwards in time to the same level of difficulty experienced by the first community. Rather, the Qur'anic guidelines should lead the various communities towards progressive change within the context of universal Qur'anic guidance. To restrict future communities to the social shortcomings of any single community—even the original community of Islam— would be a severe limitation of that guidance.

I propose that the way to believe in 'the whole of the book' (3:119) is to recognize that 'spirit' of the book and accept its worldview, vision, and ultimate intent. In examining the Qur'an, we need 'to accurately determine the rationales behind its statements, comments and injunctions'.[41] 'Even when the reason for a certain command is not explicitly stated, it is not difficult to guess it.'[42] Thus, to arrive at that ultimate intent requires the same level of pure commitment, devotion, and intellectual striving as the members of the earliest community. However, in the context of a technologically advanced world, such a commitment will require a broader, global perspective activated in, but not limited to, one's local context.

Although in some instances the Qur'an proposed immediate abolition of certain ill practices, most of the time it advocated gradual reform. Few reforms were completely implemented before the final revelation. 'If all these customs had been entirely abolished by God, several problems would have ensued ..., not many of His commandments would have been obeyed'.[43] However, the means for completing others were provided for by the flexibility and intent of the text itself. 'It was considered wise ... not to totally abolish some of the reprehensible traditions such as polygamy [or slavery], as there were so many difficulties involved.'[44]

In particular, some reformation which benefited the status of women was allowed to progress slowly because of the contexts in which these changes were to take place.

God's permissibility only showed man's cruel heart, his inability to submit to truth and justice, and his immoral character, acquired from the worst pre-Islamic customs.... Had it not been for the viciousness in his mind, his misguided soul and cruel heart, God would not have granted him then such allowances that He disliked and which were meant to vanish with time.[45]

It is continuing that natural evolution in society which is indicated in the Qur'an that explains why many Muslim countries have instituted further legal and social reforms with regard to women. These reforms operate outside the literal content of some Qur'anic passages and make modifications on the basis of greater Qur'anic intent with respect to such issues as repudiation, polygamy, inheritance, and the rules for witnessing, etc.

Polygamy

Many Muslim nations which now consider polygamy unconstitutional have justified such changes in legislation on the basis of the overall Qur'anic perspective on marriage, as well as on modern Islamic perspectives of marriage. The marriage of subjugation at the time of revelation was premised on the need for females to be materially provided for by some male. The ideal male for a female child was the father, and for the adult female, the husband. This economic perspective of marriage—as indicated by several verses discussed above—will also be reviewed here with regard to polygamy.

If you fear that you will not deal justly with the orphans, marry women of your choice, two, three or four. But *if you fear that you will not be able to do justly (with them), then only one*, ...to prevent you from doing justice. (4:3).

First, this verse is about treatment of orphans. Some male guardians, responsible for managing the wealth of orphaned female children, were unable to refrain from unjust management of that wealth (4:2). One solution suggested to prevent mismanagement was marriage to the female orphans. On the one hand, the Qur'an limited this number to four, and on the other hand, the economic responsibilty of maintaining the wife would counterbalance the access to the wealth of the orphaned female through the responsibility of management. However, most proponents of polygamy seldom discuss it within the context of just treatment of orphans.

In fact, as far as they are concerned, the only measurement of justice between wives is material: can a man equally support more than one wife? This is an extension of the archaic idea of marriages of subjugation, because fairness is not based on quality of time, equality in terms of affection, or on spiritual, moral, and intellectual support. These general terms of social justice are not considered with regard to just treatment with wives.

It is especially clear that this verse is concerned with **justice**: dealing justly, managing funds justly, justice to the orphans, and justice to the wives, etc. Justice is the focus of most modern commentaries concerned with polygamy. In the light of verse 4:129—'You are never able to be just and fair as between women ...' —many commentators assert that monogamy is the preferred marital arrangement of the Qur'an. Surely, it is impossible to attain the Qur'anic ideal with regard to mutuality ('They [feminine plural] are raiment for you [masculine plural] and you are raiment for them' (2:187)), and with regard to building between them 'love and mercy' (30:21), when the husband-father is split between more than one family.

Finally, with regard to three common justifications given for polygamy, there is no direct sanction in the Qur'an. One is financial: in the context of economic problems such as unemployment, a financially capable man should care for more than one wife. Again, this assumes that all women are financial burdens: reproducers, but not producers. In today's world a lot of women neither have nor need male supporters. For one thing, it is no longer accepted that only men can work, do work, or are the most productive workers, in all circumstances. With regard to work outside the home, i.e. paid employment, the market is based on productivity. Productivity in turn is based on a number of factors, and gender is only one of them. Surely, polygamy is no simple solution to complex economic problems.

Another rationale given for a man having more than one wife centres on the woman who is unable to have children. Again, there is no mention of this as a rationale for polyygamy in the Qur'an. However, the desire for children is natural. Thus, consideration for the barren man and the barren woman should not exclude either from the chance of marriage, nor from the care and upbringing of children. What possible solution is mutually available when the wife or husband is sterile and the couple cannot have their own children?

In a world of war and devastation, there are still orphaned Muslim (and other) children who would benefit from the love and care of childless couples. Perhaps caring for all of the earth's children might be practised by Muslims in the light of global catastrophes still unresolved. One's own blood relations are important, but perhaps not in terms of the final judgement of one's ability to care and nurture.

Finally, the third rationale given for polygamy not only has no sanction in the Qur'an, but is clearly un-Qur'anic as it attempts to sanction men's unbridled lust: that if a man's sexual needs cannot be satisfied by one wife, he should have two. Presumably, if his lust is greater than that, he should have three, and on until he has four. Only after this fourth are the Qur'anic principles of self-constraint, modesty, and fidelity finally to be exercised.

As self-constraint and fidelity are required at the onset for the wife, these moral virtues are equally significant for the husband. It is clear that the Qur'an does not stress a high, civilized level for women while leaving men to interact with others at the basest level. Otherwise, the mutual responsibility of *khilafah* (trusteeship) would be left to one half of humanity while the other half remains near the animal state.

Witness

Another consideration in contemporary discussion on the issue of women in the Qur'an focuses on the woman as a potential witness. Is one male witness equal to two female witnesses and, therefore, one male equivalent to, or as good as, two females absolutely? In the related verse, a 'record in writing ...' 'when a debt is contracted' calls for two witnesses, 'if two men be (not at hand) then a man and two women, of such as you approve as witnesses, so that if the one errs (*tudilla*) the other can remind her ...' (2:282). In the wording of this verse, both women are not called as witnesses. One woman is designated to 'remind' the other: she acts as corroborator. Although the women are two, they each function differently.

In addition, there are some contextual considerations regarding the need for more than a single witness in the first place. The purpose is to see to it that there is no error—intentionally or unintentionally—regarding the terms of the contract. Fazlur Rahman takes exception to the application of the literal wording of the verse in all future transactions as follows:

...since the testimony of a woman being considered of less value than that of a man was dependent upon her weaker power of memory concerning financial matters, when women become conversant with such matters–with which there is not only nothing wrong but which is for the betterment of society—their evidence can equal that of men.[46]

Thus the verse is significant to a particular circumstance which can and has become obsolete. However, I have found no commentary which considers the matter of intentional error. The call for two witnesses 'of such as you approve as witnesses' in the first place indicates an attempt to prevent corruption. If one goes wrong, or is persuaded to give untrue testimony, the other is there to support the terms of the contract. However, considering that women could be coerced in that society, if one witness was female, she would be easy prey for some male who wanted to force her to disclaim her testimony. When there are two women, they can support each other—especially in view of the term chosen: if she (*tudilla*) 'goes astray', the other can (*tudhakkira*) 'remind' her, or 're-call her attention' to the terms of agreement. The single unit which comprises two women with distinct functions not only gives each woman significant individual worth, but also forms a united front against the other witness.

In addition, one male witness plus this two-female unit, does not equate to a two-for-one formula because otherwise, four female witnesses could replace two male witnesses. Yet, the Qur'an does not provide this alternative. Despite the social constraints, at the time of revelation—inexperience and coercion of women—a woman was nevertheless considered a potential witness.

Even at that time of severe social, financial, and experiential constraints, the Qur'an recognized the potential of women's resources. In this modern era, such revolutionary consideration of women's potential should lead to greater promotion of her contributions to a just and moral social system, and end exploitation of her and others in society. Such a social system can be attained only through the encouragement of learning and experience for the male and the female.

Finally, the consideration regarding witnesses in this verse is specific to certain types of financial contracts. It was not meant to be applied as a general rule. Whenever the Qur'an does not specify gender in terms of a witness, androcentric interpretation concludes it to mean the male witness, exclusively. As such, the formula applied requires twice the number of females—not only for witnessing but for other aspects of woman's participation as well.

This limitation regarding financial transactions does not apply in other matters. The call for two women and one man for witnessing financial contracts is not a general rule for women's participation, nor even for all witnessing. Other requests for witnesses should be for unspecified gender. Therefore, anyone deemed capable of witnessing has the right to be one.

Inheritance

The mathematical formula of two-to-one has been—albeit erroneously—reinforced through oversimplification of the Qur'anic discussions regarding inheritance. Although the initial Qur'anic statement, in verses 4:11-12, makes 'the share of the male . . . equivalent to the portion of two female (siblings)', a complete look at this same verse enumerates a variety of proportional divisions between males and females. In fact, if there is one female child, her share is half the inheritance. In addition, the consideration of parents, siblings, distant relatives, as well as offspring is discussed in a variety of different combinations to indicate that the proportion for the female of one-half the proportion for the male is not the sole mode of property division, but *one* of several proportional arrangements possible.

Such variety of possibilities emphasizes two points with regard to inheritance: **1**. In no way are females, including distant ones, to be disinherited. This is especially true for the pre-Islamic customs still prevailing today, which give the inheritance of even female offspring to some male relative, no matter how distant. **2**. All distribution of the inheritance between the remaining relatives must be equitable. According to these verses, such 'equity' in distribution of inheritance must take the actual *naf'a* (benefit) of the bereft into consideration.

The full extent of the Qur'anic provision requires a look at other details which can lead to a redistribution of the inheritance according to the circumstances of the deceased and of those who inherit. The division of inheritance requires a look at all of the members, combinations and benefits. For example, if in a family of a son and two daughters, a widowed mother is cared for and supported by one of her daughters, why should the son receive a larger share? This might not be the decision if we look at the actual *naf'a* of those particular offspring.

The Qur'an does not elaborate all possibilities. However, by providing a variety of scenarios, it is clear enough that many combinations can and do exist, which must be considered for the equitable distribution of inheritance.

Finally, one-third of the wealth can be bequeathed, with no restrictions upon whom the beneficiary is and without decreasing the divisions of the remaining wealth. In fact, the proposal is quite flexible provided, of course, equity is met.

In short, the matter of inheritance involves these considerations; **1**. distribution to both male and female surviving relatives; **2**. some wealth can be bequeathed; **3**. consideration must be given to the circumstance of the bereft, their benefit to the deceased, and the benefits of the wealth inherited.

Male Authority

Although the basis of Qur'anic discussion on society was particular to the existing system, it also gives general principles from which to derive solutions to social problems in other contexts. With regard to authority, the prevailing attitudes were patriarchal. As with other matters in society, the Qur'anic solutions to social problems reflect the prevailing attitudes in ancient Arabia.

The general principle for leadership in the Qur'an is similar to the rule for fulfilling any task, that it should be filled by the one 'best suited'. That person is the one best suited on the basis of whatever qualifications or characteristics are necessary to fulfil that task: biology, psychology, education, finance, experience, etc. This principle works in a number of complex social arrangements: the family, society at large and leadership.

With regard to leadership, the ancient (and modern) Arabian patriarchy yields certain advantages for men. Certainly where males had public privileges, experiences, and other advantages, they were best suited to operate in the political and financial arena. It was erroneously concluded that men would always have the advantages that would make them most suitable for leadership. Yet these advantages were not restricted to men in the Qur'an. Provided the woman has the motivation, opportunities should be made available. Her capacities ... to become 'best suited' for a number of tasks not common fourteen centuries ago should have greatly increased by now.

Despite this difference in opportunities between men and

women, even at the time of revelation, there is nothing implied or stated in the Qur'an which supports the opinion that males are natural leaders. Even in the context of patriarchal Arabia, the Qur'an gives the example of a female ruler. As discussed above, Bilqis is depicted extremely well in the Qur'an.[47] In fact, other than the prophets, she is the *only* ruler in the Qur'an who is given favourable consideration. The Qur'an refers to her characteristics of wisdom and independence as a leader.

The Qur'an does not restrict the female from being in authority, either over other women or over both women and men. However, there is the implication that the Qur'an inclines towards seeing necessary tasks fulfilled in society in the most efficient manner. Neither male nor female will be equally beneficial in every situation. To force even modern patriarchal societies to submit before a female ruler would be detrimental to the harmonious welfare of that society.

However, choosing the one best suited for the tasks at hand is a dynamic process. Continual assessment of a current situation should yield sufficient information about the qualification for fulfilling any task. A more independent and insightful woman might better lead a people into their future endeavours. Similarly, a husband may be more patient with children. If not predominantly, then perhaps temporarily, such as when the wife falls ill, he should be allowed to fulfil that task. Just as leadership is not an eternal characteristic of all males, child caring is not an eternal characteristic of all females.

Child Care

Again the prevailing system in many societies has determined that the responsibility of caring for the children is appropriate to the female. The convenience of this in many families and with many women who have a 'feminine' temperment inclined towards nurturing has reinforced this man-made determination as if it were inherent.

The Qur'an gives rights to both parents with regard to the children and equally allows for their emotional attachment: 'A mother should not be made to suffer because of her child, nor should he to whom the child is born (be made to suffer) because of his child' (2:223). It allows for the mother to suckle a child, but leaves it to her own discretion: 'Mothers shall suckle their children . . . (that is) for those who wish to complete the suckling' (2:223). If instead a couple or a mother decides 'to give [the] children out to nurse, it is no sin for you' (2:223). Thus the basic nurturing of even the very young child is optional.

Yet, the tendency has always been to attach all forms of child care—and in addition all forms of housework—to the woman. Although this division of labour suits some families, especially when the father is working outside the home and provides materially for his family, it is, nevertheless, only one solution and does not have explicit Qur'anic ordinance.

In families where both husband and wife are providing for the material maintenance of the family, it is an unfair burden for the woman to be solely charged with all housework. If she has managed to increase her good deeds, then there are similar opportunities for the man to increase his by participating more in the housework and child care. In addition, this demonstrates to the children the ultimate Qur'anic system of evaluation which does not specify the deed: 'Whoever does good, whether male or female and is a believer, all such will enter Paradise' (4:124).

Such a flexible, integrated, and dynamic system of co-operation benefits a multiplicity of societies and family types. It places proper recognition on the single-parent households of today. If it is headed by a woman, then full responsibility, authority, and qiwamah are placed on her shoulders. Similarly, daughters may also assume the responsibilities of extended relatives and disabled husbands.

When the Qur'an is viewed in its entirety and not atomistically, the concern for the role of woman in society and the potential she has, would necessarily be broadened from the demeaning and meaningless existence which renders her no greater than a procreating animal able to function only as a domestic servant. Likewise, her skills and participation on the home front would be acknowledged as significant and meaningful, such that all who aim at performing good deeds would rush to fulfil the same tasks.

If the aim of Islamic society is to fulfil the intentions of the

Qur'an with regard to the rights, responsibilities, potentialities, and capacities of all its earnest members, then those who truly believe in the Qur'an would equally wish for the woman the opportunities for growth and productivity which they demand for the man. The man would equally be charged with nurturing and caring for the family over and above attending to its material concerns. In short, both would be well-rounded and mutually beneficial to each other, to the family, and therefore to society at large. Only then is the true potential of the *khilafah* cultivated. The family acts as the initial arena of practice. Surely, as the Prophet says, 'The best of you is he who is best to his family. . . .'

---

[1]See, for example, S. H. A. L. Shamma, *The Ethical System Underlying the Qur'an* (Tubingen: Hopfer Verlag, 1959), which discusses each ethical principle in the Qur'an with regard to Qur'anic chronology.

[2]It should not be overlooked in my literal interpretation, that 'the wombs that bore you' is also used as a metaphor for the blood ties of family relations in general.

[3]To further substantiate this point of view, see the discussion of Mary, Mother of Jesus, in Chapter 2.

[4]Which I will discuss in detail below.

[5]There is a distinction between *wahy*: receiving divine communication, and *risalah*: receiving divine communication concerning the destiny of humankind which includes the obligation to transmit the information of that *wahy* to humankind at large.

[6]Maryam and Umm-Musa among the women.

[7]See, for example, verse 22:75.

[8]'See how We prefer one above another, and verily the Hereafter will be greater in *darajat* and greater in preference [*tafdil*]' (17:21).

[9]Carol Tarvis and Carole Wade, *The Longest War: Sex Differences in Perspective*, 2nd edn. (Orlando: Harcourt Brace Jovanovich, 1984), p. 3.

[10]Ibid., p. 20.

[11]This is one of the verses which demonstrates that an Islamic society allows for economic classes.

[12]In addition, the preceding verses, 221–37 discuss at length other details related to marriage, divorce, and widowhood.

[13]Edward William Lane, *An Arabic-English Lexicon*, 8 vols. (London: Librairie du Libon, 1980), Part 5, p. 2017.

[14]As defined by Milton J. Cowan, *A Dictionary of Modern Written Arabic*, 3rd edn., (ed.) Hans Wehr (Ithaca, New York: Spoken Language Services, Inc., 1976).

[15]For example, the Children of Israel were preferred over 'other creatures' (the same term has been translated in Chapter 1, the Fatihah, as 'the worlds') in verses 2:47, 2:122, and 7:40. This preference is usually understood to mean that they were chosen to receive Prophets and the revelations.

[16]This is the *ba al-sababiyyah* known in Arabic as the *bi* for a reason or purpose. It establishes a conditional relationship between two parts of the sentence or clause. The first part is conditional upon, and cannot be attributed without the second part.

[17]Aziza al-Hibri, 'A Study of Islamic Herstory: Or How did We ever get into this Mess?', *Women and Islam: Women's Studies International Forum Magazines*, 5 (1982); p. 218.

[18]Which I will discuss in detail below.

[19]Aziza al-Hibri, op. cit., pp. 217-18.

[20]I have translated *ba'd* in its usual meaning of 'some' or 'a portion of'. However, there is also usage of *ba'd* + masculine plural pronoun + *ba'd* which means 'each other' with no particular number or gender implied. In other words, a degree of vagueness surrounds this statement. It could also mean women have a preference over men.

[21]Al-Hibri, op. cit., p. 218.

[22]Vol. I, p. 523.

[23]Vol. II, p. 117.

[24]'Islamic Herstory,' p. 217.

[25]Ibid., p. 218.

[26]For example, I would include Pickthall because he translates this passage as unrestricted 'Men are in charge of women'. Al-Zamakhshari, in *al-Kashshaf an Haqa'iq al-Tanzil wa 'uyun al'Aqawil fi wujud al-Ta'wil*, 4 vols. (Bayrut: Dar al-Ma'arif, n.d.), Vol I, p. 523, states the terms he believes of Allah's preference of men over women. Abbas Mahmud Al-'Aqqad, *al-Mar'ah fi al-Qur'an* (Cairo: Dar al-Hilal, 1962) p. 7 states the same. Finally, Maududi interprets it this way.

[27]Vol. II, pp. 648-53.

[28]Sayyid Qubt, *Fi-Zilal al-Qur'an*, 6 vols. (Cairo: Dar al-Shuruq, 1980), Vol. II, p. 650.

[29]Ibid.

[30]Ibid.

[31]It is also extended to other than humans (2:116, 30:27).

[32]Ibid., p. 652.

[33]Part 8, p. 2795. Although he distinguishes between the *nushuz* of the wife and the husband, he also defines it as the one 'hates or dislikes' the other and is an 'evil companion' to the other.

[34]Qutb, Vol. II, p. 653.

[35]Ibid.

[36]The verb here is *bagha* which Pickthall has translated elsewhere as 'doing wrong against someone or something' (verses: 10:23, 42:42, 49:9). Lane's

translation, Part 1, p. 231, also suggests acting wrongfully or injuriously towards another person, oppressing or seeking to hurt another.

[37]See, for example, Abdullah: al-Na'im's *Towards an Islamic Transformation: Civil Liberties, Human Rights and International Law* (New York: Syracuse University Press, 1990, in which he discusses a methodology for applying general principles of the Qur'an over limited, specific statements from the Madinan period. See also Nazirah Zein-ed-Din, 'Removing the Veil and Veiling', translated from the 1928 edn. by Salah-Dine Hammoud for *Women and Islam: Women's Studies International Form Magazine*, 5 (1981): 221-7.

[38]Maria Riley, *Transforming Feminism* (New York: 1989), p. 102

[39]This author does not distinguish between the word of God and Islam, but this is an essential distinction for this research.

[40]Aziza al-Hibri, 'Islamic Herstory,' p. 207.

[41]Rahman, *Islam and Modernity: Transformation of an Intellectual Tradition*, (Chicago: The University of Chicago Press, 1981), p. 19.

[42]Ibid., p. 18.

[43]Nazirah Zein ed-Din, see footnote 37, p. 224.

[44]Shaykh Al-Ghalaynini, *Al-Islam Ruhu al-Madaniyyah*, quoted by Nazirah Zein-ed-Din, 'Removing the Veil and Veiling', p. 223.

[45]Nazirah Zein-ed-Din, op. cit., p. 223.

[46]Rahman, *Major Themes in the Qur'an* (Chicago and Minneapolis: 1982) p. 49.

[47]See Chapter 2 for a detailed discussion of the Qur'an's treatment of Bilqis.

Conclusion

*T*HE purpose behind this research was to consider some aspects of gender equality in the Qur'anic *Weltanschauung*. In this conclusion I will explore the results of this research and explain the benefits of the methodology and perspective of my research.

Interpretation of the Qur'an can be divided into two levels: reading and exegesis. At the level of a 'reading', every reader interprets while in the act of reading. This level of interpretation is shaped by the attitudes, experiences, memory, and perspectives on language of each reader: the prior text. With regard to most topics, that prior text reflects certain perceptions of gender. The specific concerns with gender in the text bring such perceptions to the foreground and focus on them in particular terms. At the level of exegesis, an attempt is made to overcome individual perspectives to approach the Qur'an more objectively. Thus, the use of several supportive skills—Qur'anic 'sciences'—and the application of hermeneutical methods are intended to lead towards that objectivity. However, no exegete is able to remove the significance of the personal reading and the force of the prior text from interpretation.

The existence of so many exegetical works (*tafasir*) indicates that, with regard to the Qur'an, the interpretation process has existed, and will probably continue to exist, in a variety of forms. It is essential that the natural adaptive nature of interpretation, from individual to individual and from time and place to time and place, should continue unabated until the end of time—on the one hand, because it is natural, and, on the other hand, because only through continued interpretation can the wisdom of the Qur'an be effectively implemented. This implementation will be specific to the varying experiences of human civilization.

No interpretation is definitive. I have attempted here to render a reasonably plausible interpretation to some difficult matters. The basis for this plausibility is the significance I draw from the text with regard to the modern woman: the significance of her life-style to her concerns and interactions in her context. I am also influenced by prior text. I have demonstrated the relevance of the Qur'an to the concerns of the modern woman. In doing so, I provide a reading that transcends some of the limitations in previous interpretations. On one hand, some limitations exist in the text—such as when it specifically addresses the social situation in Arabia at the time of revelation—on the other hand, most limitations are reflections of the interpreters who restrict the universality of the divine message to their individual perceptions. It is failure to understand this disparity between particular usages in the Qur'an and its general usages that have led to some of the variations in opinions concerning the overall Qur'anic world-view.

I believe the Qur'an adapts to the context of the modern woman as smoothly as it adapted to the original Muslim community fourteen centuries ago. This adaptation can be demonstrated if the text is interpreted with her in mind, thus indicating the universality of the text. Any interpretations which narrowly apply the Qur'anic guidelines only to literal mimics of the original community do an injustice to the text. No community will ever be exactly like another. Therefore, no community can be a duplicate of that original community. The Qur'an never states this as the goal. Rather, the goal has been to emulate certain key principles of human development: justice, equity, harmony, moral responsibility, spiritual awareness, and development. Where these general characteristics exist, whether in the first Muslim community or in present and future communities, the goal of the Qur'an for society has been reached.

The attitude of the individual exegete has helped to shape the interpretation he or she gives to specific verses and general Qur'anic principles. That most of the *tafasir* (exegetical works) are done by males indicates something about women and interpretation. Perhaps women can agree and have agreed with these, in total or in parts. Perhaps women can disagree and have disagreed, but willingly acquiesced to these 'authorities', or were unable to substantiate their disagreement. Whatever the case, no record exists, in our Islamic legacy, of a meaningful discourse between the preceptions, experiences and reflections of women and men about both their different and similar understandings of the text.

A significant part of the motivation behind my consideration of this subject was to challenge some of the attitudes and the resulting interpretations given with regard to the subject of woman and the Qur'an. I explicitly challenge the arrogance of those men who require a level of human dignity and respect for themselves while denying that level to another human, for whatever reason— including simply because she is a woman. In particular, I reject the false justification of such arrogance through narrow interpretations or misinterpretations of the Qur'anic text, namely interpretations which ignore the basic social principles of justice, equality, and common humanity. As one writer suggests, it is only through religion 'and in this case, the explicit text of the religion of Islam' that men can ever hope to overcome a cruelty of nature which has led them to commit acts of oppression or abuse incomparable to any acts performed by women.[1]

Is this then the inherently proper character with which a man is privileged over woman? If a physically weak but proper and intelligent individual happens to be humiliated and abused by another physically strong but brutal individual, should the latter be considered wiser than the former?[2]

If readers of the Qur'an have assumed in any manner that men are superior to women intellectually, spiritually, ontologically, etc.; that men are 'in charge of women'; that men have a more significant role in the continuation of society; that men are natural leaders; that men should 'rule' the family and get obedience from women; that women do not have to participate and contribute in order to maintain the family and society or that her participation is marginal; then those readers will interpret the Qur'an in accordance with those assumptions.

Sometimes the open-endedness of the Qur'an, which does not give explicit social functions or attribute explicit values to those functions, has been interpreted to mean that the Qur'an supports existing systems of evaluation—whether at the time of the Prophet or in subsequent Muslim communities. In the light of this, Muslims have used verses from the Qur'an both to support a woman's subservience to a man and to defend her rights of equality and independence.

Although my goal was to demonstrate the significance of the greater Qur'anic concepts and principles, it was also necessary for me to examine—sometimes in great detail—the meanings and implications of individual words or word-groups. Words, particles, syntax of verses and their contexts are all part of the multi-layered interpretation process. Some Qur'anic words have been used to restrict the application of greater Qur'anic concepts and principles. Thus, words in the Qur'an must be examined in the light of the overall Qur'anic *Weltanschauung*.[3] The words used in the Qur'an were part of the existing language and culture of seventh-century Arabia. Yet, sometimes the Qur'an incorporated them in a different manner to reflect its greater intent and purpose of guidance to humanity.

It is important to focus on careless and restrictive interpretations of words, verses, and contexts, especially when these have damaging consequences on any individuals. In my research, I was particularly concerned with the consequences on women. The lives of women have been comprehensively, adversely affected by interpretations.

Rather than be elevated by the text to transcend their own limitations, some interpreters bring the text down to their level when they project narrow or negative meanings which suit their individual whims, perceptions, and prejudices about women. This is most often done on the basis of a single word! Some of these words do not even exist in the text. Other negative terms, if used at all, in the Qur'an, are neither directly nor exclusively associated with women. Even when a negative word is coincidentally used exclusively with reference to women, it does not mean that *all* women necessarily fall prey to the indications of that word, nor that men are exempted from falling prey or permitted to fall prey. The interpretations of these words and other syntactical structures have not been juxtaposed with the entire Qur'anic world-view.

Tabbaruj, (wanton display), in verses 33:33 and 24:60, for example, is used only with reference to women. 'O wives of the Prophet . . . stay quietly in your houses and make not a wanton display like that of the former Time of Ignorance....'(33:32-3). The Qur'an orders women (precisely the wives of the Prophet) not to *tabbaruj* like the *tabbaruj* of the days of Jahiliyyah. And 'Older women . . . can lay aside their (outer) garmets, provided they do not make a wanton display of their beauty ...' (24:60).

Although only women are addressed in each passage, it does not mean that the women being addressed were guilty of *tabbaruj*, nor that women are exclusively prone to it. Rather, it expresses that the 'wanton display' (perhaps of some, but in any case only) during the Times of Ignorance, or Jahiliyyah, is prohibited.

Is this Qur'anic prohibition applied *only to women* because they were specifically addressed—implying therefore that it is all right for men to practise it? This is clearly not so. Because women were mentioned *consequently* in a verse, it is no indication that the verse must be restricted in its application to only that gender. All who wish to attain moral excellence in the Qur'anic world-view of social behaviour must strive to avoid going out of the house with the express purpose of wanton display. Such a gender-constraint on the understanding and application of the verse is unnecessary. Rather, the women are used to state a general social principle.

Incidentally, verse 33:33 has been used to make the general rule that women are restricted from going out of their houses altogether! Instead of stressing the limitation of the 'going out' for the purpose of wanton display, it has been used to prohibit **all** going out. Women must remain indoors 'unless absolutely necessary'. The specific characteristic of going out for wanton display that is prohibited is made into a general rule for **all** going out. Meanwhile, the general rule for all members to avoid going out for 'wanton display' is restricted to women.

One author, Al-'Aqqad, in his book *al-Mar'ah fi al-Qur'an* (Woman in the Qur'an),[4] has a particular concern for seeking out or creating negative attributes to women and projecting them as Qur'anic portrayals of the nature of woman. The inaccuracy and consequent destructiveness of analyses such as this needs to be pointed out. For example, he says that when *kayd* (planning a strategy against) is used in connection with women to indicate a negative attitude that they are excessively prone to.[5] A simple review of the Qur'anic index will reveal that this word is used with males (3:120, 52:46, 100:2), with females (12:33, 34, 50), and even with Allah (7:84, 68: 45, 56:16). In another example, he associates a word with the weak, particularly women—which does not even occur in the Qur'an.[6] He has clearly extended his perception of women on to the text.

Such associations between personal perceptions and Qur'anic concepts need critical review. When a Muslim thinker claims authority to render interpretations of the text, most Muslims who read these interpretations will assume that they are legitimate aids to understanding. The effects of such projections from 'authorities' on to the Qur'anic text foster a witch-hunting mentality which looks for 'inherent' evil in woman, and then justifies constraints on her every move.

The Qur'an clearly rejects any such notion of the 'inherent' evil of woman. It explicitly demands respect for her 'inherent' good as potential child-bearer (and primary nurturer). It places her on absolute par with man in terms of the spiritual potential (to know and serve Allah) and the potential to attain Paradise, provided she and he strive to realize such potential:

For Muslim men and Muslim women, for believing men and believing women, for devout men and devout women (i.e. who have *qunut*), for true men and true women, for men and women who are patient and constant, for men and women who humble themselves, for men and women who give in charity, for men and women who fast, for men and women who guard their chastity, and for men and women who engage much in Allah's praise—for them has Allah prepared forgiveness and great reward.(33:35).

Some of the greatest restrictions on women, causing them much harm, have resulted from interpreting Qur'anic solutions for particular problems as if they were universal principles. For example, I have discussed at length how the interpretations of verse 4:34 have been used to indicate that women are inferior to men, that a woman must obey her husband, and that if she does not, he can beat her.[7]

In addition, I have demonstrated that although there are Qur'anic references which mention women, they are usually not meant exclusively for women. Verses in *Surat al-Tahrim* (66), which are used to give examples of 'good' and 'bad' women, are excellent examples. As the verses conclude, Maryam, one of the 'good' women, is classified among the *qanitin*. That is, she is among the category of *people* who are characterized by a co-operative nature devoted to Allah. The use of the masculine plural form demonstrates the extension of this human model to all 'those who believe'—as the verses state explicitly—rather than to women alone. If the Qur'an wanted to restrict the verse to women only as is so often implied, it would have simply used the feminine plural form, *qanitat*.

In addition, on purely grammatical grounds, I extend the meaning of verses which use the masculine plural form to apply equally to females. Men and women are meaningful to each other. When the Qur'an says '*ba'du-kum min ba'd* (you proceed from each other)',[8] it means the male and the female are of the same human status.[9] I extend this meaning to other messages encoded in the Qur'an. Whether the particular references made are masculine or feminine, the principles embodied in the message are unrestricted. Good and evil are non-gender specific. In this we see that

On the ethico-religious level, the position of men and women are on equal standing, both as to their religious obligations towards God and their peers as well as the consequent reward or punishment.

The moral and religious equality of the sexes before God represents the highest expression of the value of equality. Furthermore, the ethico-religious equality of women is independent of and not subject to, change of social situation. This value, then, enjoys a higher degree of priority over any value that is dependent upon a changing social context.[10]

Some of the problems I have considered here, I attempt to resolve by applying Qur'anic intent, as understood from a holistic review of its ethico-moral statements concerning society. It is this method that is most useful in adapting the text to a multitude of culturally diverse situations in a constantly changing world of social communities.[11]

The most important point I raised, with regard to the Qur'an's approach to the subject of woman, is that the attitudes towards women at the time and place of the revelation helped to shape the particular expressions in the Qur'an. The concerns it addressed were particular to that circumstance. These particulars are not meant as the entire Qur'anic intent, but are the means for determining that intent provided we have appropriate historical information. For example, the reverential addresses to women, in the Madinan period, are commensurate with the style practised at that time. The Qur'an uses that style to indicate reverence—not to indicate that it is the *only* means to show reverence. Therefore the use of that same style can be adopted, but there is no indication that other styles are negated.

Therefore, I focused a little on the chronological development of the Qur'anic programme for women (and indeed towards the development of all·social reforms) because such a focus reveals the Qur'an's natural progression in overcoming deep rooted attitudes and the actions that result from them.[12] I hope this will be applied when new communities attempt to develop within an Islamic framework. Only gradually will they reach the goal they seek and approach an Islamic ideal.

With regard to chronology, it has always been pointed out that the Qur'an brought many changes to the status of women and for their consideration in society. 'Umar Ibn al-Khattab was quoted as saying: 'By God, we did not use to pay attention to women in Jahiliyyah until God said about them in the Qur'an what is said, and gave them their share in matters.'[13] The continued change which the Qur'an put into motion was not meant to stop when the revelation was completed.

Some important social changes not completely instituted by the end revelation (like the total abolition of slavery), were given enough clear indications of the direction that the Qur'an intended. It therefore seems unbelievable that it was only so many centuries after the revelation that they were implemented.

In addition, by extending certain Qur'anic social principles, Muslim communities should have evolved into leading examples of humane and just social systems. Certainly, the Qur'an never intended for institutions like slavery to continue even though it never explicitly prohibited it. There are unlimited possibilities for reform within the parameters of equity, mutual honour, respect, and consultation enunciated by the Qur'an.

Thus I have extended the Qur'anic principles of social justice, equity, mutual honour and moral responsibility, into the context of the present era and at least given some thought to the evolution of those principles in the future. My specific goal was to demonstrate the adaptability of the Qur'anic world-view to the issues and concerns of women in the modern context. If the attitudes of women and men towards each other had not progressed in the last 1,400 years, we would be in a pitiful state.

Muslim scholars abstractedly agree that such principles as discussed here are indeed in the Qur'an, but they interpret and apply them differently. For example, 'equal' and 'equality' can be defined in more than one way. If the meaning of 'women are equal to men' is that men and women are the *same*, I have argued that the Qur'an reveals a necessary and complementary distinction between men and women. However, I have shown this to mean that they have the same rights and obligations on the ethico–religious level, and have equally significant responsibilities on the social–functional level.

I further argue that the existence of this equality on the ethico–religious level is more significant than the different values that various social systems have attributed to the social–functional level which have led to the absence of equality. The Qur'an offsets tendencies within social systems towards different values by assigning equitable compensation for the deeds performed by each individual in their social–functional context. I believe that my definition unambiguously exemplifies the extension of the overall Qur'anic human rights equally to women and men.

My goal has not been to defend a definition of equality between man and woman which continues or reverses the very limited androcentric world-view such that what men do is still used as the model upon which social worth and justice are understood. Sometimes, there are significant lessons to be learned from the nurturing and accommodating experiences of women, to say nothing of their fidelity. Such was surely the case with the Qur'anic account of Bilqis as a ruler. Instead of using her model as a universal that all leaders should emulate, androcentric interpretation says a woman cannot be a leader! And yet such lessons are significant in the development of the true *khalifah* (trustee) on earth.

Then again, men and women are not so equal that the differences between them—which help to create harmony in their inter-relation—should be obliterated. I believe this is spiritually, morally, and socially counterproductive. The Qur'anic evidence substantiates this view and stresses the significance of each to the other.

With regard to social justice, it becomes necessary to challenge patriarchy—not for matriarchy, but for an efficient co-operative and egalitarian system which allows and encourages the maximum participation of each member of society. This system would truly respect each gender in its contributions, and all tasks that are contributed. This would allow for the growth and expansion of the individual and consequently for society at large. As such, women would have full access to economic, intellectual, and political participation, and men would value and therefore participate fully in home and child care for a more balanced and fair society.

Another Qur'anic principle which I extend to include the relations between man and woman is mutual honour and consultation (*shura*). Normally, *shura* is used to discuss and resolve problems between responsible adults. Yet, too many authors have denied this to husbands and wives in conflict. The Qur'an encourages men and women to marry as a safeguard of moral behaviour between the sexes. However, some interpretations of the rights and responsibilities between the married couple have so severely restricted woman that marriage becomes an institution of oppression for her. If marriage is the means by which a woman is stripped of her individuality and her self-respect as a human equal in humanity and in spiritual capacity to any man, then this is clearly against the Qur'anic intent for a just and moral social order that encourages the doing of good and forbids the doing of evil. It is necessary to resolve this flaw of interpretation.

The rights and responsibilities of women in the twentieth century, as well as social, political and economic opportunities for their full participation which have been discussed and implemented at the policies level, are reinforced by the Qur'anic reading I discuss. It is now unquestionable that woman has increased her worth to society at large. Consequently, the recompense she will receive because of her multi-faceted role would also be increased according to the Qur'anic scale of recompense.

In order for the man in these times to continue to gain additional recompense, he too must expand the dimensions of his participation within society. There is still a significant contribution he can make by caring more for the children and helping more in the home. It will, at the very least, demonstrate that wives are motivated by the same needs and wants, or inhibited by the same fears and insecurities as husbands.

The perception of the role of woman affects every aspect of the lives of Muslim men and women. It is time to remove such oppressive restrictions and encourage expansive ideals in the participation of women and men to allow for their maximal contribution to the family as well as to society at large. When the contribution of each member (whether male or female) is valued, an increase in total productivity will affect the entire community.

Across time and culture, the division of labour between men and women has exhibited some variations. Sometimes, what the men do in one society is done by women in another society. The absence of explicit Qur'anic prescriptions for dividing labour allows and supports a myriad of variations. When individual cultures decided upon or evolved divisions of labour, they had done so to achieve the ultimate advantages in their ecological context with their available human resources. Such variation similarly exists between and within individual families. Thus, mutually decided matters of contribution can be arranged within families to affect the ultimate benefits in family and society. Not all drudgery should be arbitrarily attributed to women, nor all social, political, and economic recognition be attributed to men.

Thus the Qur'anic guidance can be logically and equitably applied to the lives of humankind in whatever era, **if** the Qur'anic interpretation continues to be rendered by each generation in a manner which reflects its whole intent. This method shows its adaptability to present and future realities.

[1]Nazirah Zein-Ed-Din, 'Removing the Veil and Veiling', translated from the 1928 edn. by Salah-Dine Hammoud for *Women and Islam: Women's Studies International Forum Magazine*, 5 (1928: 221-7).

[2]Ibid., p. 223.

[3]As discussed in detail by T. Izutsu, *God and Man in the Koran: Semanties of the Kor'anic Weltanschauung* (Tokyo: The Keio Institute of Culture and Linguistic Studies, 1964).

[4]Abbas Mahmud al-'Aqqad, *al-Mar'ah fi al-Qur'an* (Cairo: Dar al-Hilal, 1962).

[5]Ibid., p. 17.

[6]Ibid., pp. 17-19.

[7]Chapter 4.

[8]See verses 2:195 and 4:25.

[9]See, for example, footnote 1 of Pickthall's translation of this verse. p. 72.

[10]John L. Esposito, *Women in Islamic Family Law* (New York: Syracuse University Press, 1982), pp. 107-8.

[11]This application of Qur'anic intent, especially to social issues, is discussed at length by Fazlur Rahman in *Islam and Modernity: Transformation of an Intellectual Tradition* (Chicago: The University of Chicago Press, 1981), pp. 1-11; and by John L. Esposito in *Women in Islamic Family Law*, pp. 106-9. Finally see Isma'il Faruqi's 'Towards a New Methodology for Qur'anic Exegesis,' *Islamic Studies*, March 1962, pp. 35-52.

[12]Sayyid Qutb wrote at length about this chronological aspect of the Qur'an with regard to social consciousness and the eventual reforms in social systems that could be reached by the Qur'anic method in his book *Ma'alam al-Tariq* (Cairo: 1965) p. 31.

[13]As quoted by Azizah al-Hibri in 'A Story of Islamic Herstory: Or How did we ever get into this Mess?' *Women and Islam: Women's Studies International Forum Magazine*, 5 (1982): 213, from *Ihya' Ulum-ad-Din* of Abu Hamid al-Afghani, Vol. II, (Cairo, Egypt: Halabi Press, 1958).

Appendix

List of Female Characters Mentioned in the Qur'an

T HERE are so few female individuals referred to specifically in the Qur'an that I am including a complete list here. The names listed here have been taken from English and Arabic exegetical works on the Qur'an. This list is prepared in three levels but with two major categories. Within each level the list is organized in the order in which they occur in the Qur'an, which does not suppose any hierarchy. These entries will include: the proposed names of these women (when available), the verses in which they have been mentioned; and a few notes about who they were or the Qur'anic reference to them.

Category I: Level 1 includes those women who have been mentioned with extremely few details about themselves. They seem to be included only to help make a story coherent, because they do not speak or perform any actions. In the Makkan period, they are mentioned in stories about particular prophets. In the Madinan period they are most often theoretical models.

Elisabeth,[1] 3:40 and 19:5,7, wife of Zakariyya.

Banati, 11:78; my daughters or girls (of my town), said by Lot.

'A'ishah,[2] 24:11, one of Muhammad's wives.

Zaynab,[3] 33:37, one of Muhammad's wives, formally the wife of his adopted son, Zayd.

banat, 33:59, Muhammad's daughters, or the girls in his town.

imra'at-Nuh, 66:10, wife of Noah.

Umm Jamil bint Harb,[4] *imra'ah*, 111:4,5, wife of Abu-Lahab.

Category I, Level 2 includes those women who perform some action, or speak some words. However, the significance of those actions and words is limited to some exclusive events in their lives and the lives of the Prophets with whom they were mentioned. In other words, these events were individual and, like Level 1 of this category, these women demonstrate no explicit religious significance. The absence of religious significance in some of these specific details confirms the humanness of the individuals and the prophets with whom they were narrated. They have been related to us in human terms involved in natural human circumstances and facing unprophetic crises in their lives and with regard to their families.

In this group are a number of women who fulfil roles usually deemed appropriate for women. Some exegetes attempt to discuss these as universal.

imra'ah Luh, 7:83, 29:33, 11:81, 15:60, 27:57, wife of Lot.

Sarah,[5] 11:71, 51:29, wife of Abraham.

Zulikhah,[6] *Imra'at-al-'Aziz*, 12:23, woman who tempts Joseph.

al-Niswa, 12:30, women guests of Zulikhah.

Raytah bint Sa'd,[7] 16:92, she who untwists the thread.

Maryam,[8] *ukht-Musa*, 20:40, 28:7, Moses' sister.

Imra'atan, 28:23, two women of Madyan.

azwaj-al-nabi, 33:28, wives of Muhammad, also known as *Ummahat al-Mu'minin*: the mothers of the Faithful.

Khawlah bint Tha'labah,[9] *al-Mujadilah*[10], 58:1, wife of Aws.

al-Muhajirat, 60:10, women refugees.

ba'd- azwaj-hu, one or more of Muhammad's wives, 66:3; in this case it is believed to refer to 'A'ishah and Hafsah.

Category II, Level 3, is a special category of women for it includes those who perform unique functions from the perspective of the book itself and from the perspective of humanity. They have performed isolated, or individual, acts of devotion, have exemplified their own religious moral commitment, and/or they form a part of an unusual circumstance which concerns all of humankind.

Eve, Hawwa', 2:35, 4:1, 7:12, *zawj* or wife of Adam.

Mary, Maryam, 3:36,37, mother of Jesus.

Hannah, Anna, or Anne,[11] 3:35, woman: *imra'ah* of Imran, mother of Mary.

'Asiyah,[12] 20:36, 28:7, 66:12, wife of Pharaoh.[13]

Umm, 20:38, 28:7, *Umm* or mother of Moses.
Bilqis,[14] *imra'ah* 27:23, Queen of Sheba.

---

[1]'Abdullah Yusuf 'Ali, *The Holy Qur'an: Text, Translation and Explanatory Notes* (Washington, DC: Islamic Centre, 1938), p. 773, footnote 2481.

[2]Ibid., p. 898, footnote 2962.

[3]Ibid., p. 1117, footnotes 3722-3.

[4]Abu al-Qasim Mahmud al-Zamakhshari, *al-Kashshaf an Haqa'iq al-Tanzil wa 'uyun al-'Aqawil fi Wujud al-Ta'wil*, 4 Vols. (Bayrut: Dar al-Ma'arif, n.d.), Vol. IV, p. 297.

[5]Yusuf 'Ali, op. cit., p. 533, footnote 1567.

[6]Ibid., p. 559, footnote 1666.

[7]Al-Zamakhshari, op. cit., Vol. II, p. 426.

[8]Ibid., p. 537.

[9]Idem, Vol. IV, p. 69.

[10]Note that the name of the chapter is *al-Mujadalah* which is the verbal noun. However, when referring to the person the above spelling is correct to indicate the use of the active participle.

[11]Yusuf 'Ali, op. cit., p. 131, footnote 375.

[12]al-Zamakhshari, op. cit., Vol. II, p. 537.

[13]It has never been shown definitively that each reference to the wife of Pharaoh is the same woman, or that the Pharaoh referred to is the same man each time. This is not significant to the analysis I will be making and for that reason, I have made them one and the same person.

[14]Yusuf 'Ali, op. cit., p. 983, footnote 3264.

Bibliography

Qur'anic Sources, Translations, Tafasir, Etc.

'Abd al-Baqi. Muhammad Fu'ad. *Al-Mu'jam al-Mufahras li Alfaz al-Qur'an al-Karim* (Istanbul: al-Maktabah al-Islamiyyah, 1983).

'Ali, 'Abdullah Yusuf, *The Holy Qur'an: Text, Translation and Explanatary Notes* (Washington, DC: Islamic Center, 1938).

_____ *The Holy Qur'an: Text, Translation and Commentary*, US edn. (Elmhurst, New York: Tahrike Tarsile Qur'an Inc., 1987).

Flügel, Gustav, *Concordance of the Koran* (Karachi: Sohail Press, 1979).

Maududi, S. Abul A'la, *The Meaning of the Qur'an*, 13 vols. 6th edn. edited by A. A. Kamal and translated by Ch. Muhammad Akbar (Lahore, Pakistan: Islamic Publication Ltd., 1983).

_____ *The Holy Qur'an: Text, Translation, Brief Notes*, translated by Muhammad Akbar Muradpuri and 'Abdul 'Aziz Kamal (Lahore, Pakistan: Islamic Publication Ltd., 1982).

Pickthall, Muhammad M., *The Glorious Qur'an: Text and Explanatory Translation* (New York: Muslim World League, 1977).

Qutb, Sayyid, *Fi-Zilal al-Qur'an*, 6 vols. (Cairo: Dar al-Shuruq, 1980).

Von Denffer, Ahmad, *'Ulum al-Qur'an: An Introduction to the Sciences of the Qur'an* (Leicester, England: The Islamic Foundation, 1983).

al-Zamakhshari, Abu al-Qasim Mahmud (467-538 AH), *al-Kashshaf 'an Haqa'iq al-Tanzil wa 'Uyun al 'Aquawil fi Wujud al-Ta'wil*, 4 vols. (Bayrut: Dar al-Ma'arif, n.d.).

Dictionaries, Lexicons, and Grammars

Cowan, J. Milton, *A Dictionary of Modern Written Arabic*, 3rd edn.

edited by Hans Wehr. (Ithaca, New York: Spoken Language Services Inc., 1976).

Hasan, Abbas, *al-Nahw al-Wafi*, 4th ed., 4 vols. (Egypt: Dar al-Ma'arif, n.d.).

Lane, Edward William, *An Arabic-English Lexicon*, 8 vols. (London: Librairie Du Liban, 1980).

The Compact Edition of the Oxford English Dictionary, 2 vols., Vol. I. A - O. (Oxford: Oxford University Press, 1987).

Muhammad, Jamil al-Din, *Lisan al-Arab*, 15 vols. (Bayrut: Dar Sadir, 1975).

Wright, W., *A Grammar of the Arabic Language*, edited and translated from the German of Caspari, 3rd edn. (Beirut: Librairie du Liban, 1974).

al-Zamakhshari, Abu al-Qasim Mahmud, *Asas al-Balaghah*, 4th edn. (Bayrut: Dar al-Tanwir al-Arabi, 1984).

Arabic Sources

'Abd-al-Wahhab, Ahmad, *I'jaz al-Nazm al Qur'ani* (Cairo: Maktab al-Gharib, 1980).

'Ali Wahbah, Tawfiq, *Dawr al-Mara'ah fi al-Mujtami'ah al-Islami* (Riyadh, Saudi Arabia: Dar al-Liwa' wa al-Tawzi', 1978).

Amin, Qasim, *Al-Mar'ah al-Jadidah* (Cairo: Matba'at al-Taraqqi, 1938).

Al-'Aqqad, 'Abbas Mahmud, *Al-Mar'ah fi al-Qur'an* (Cairo: Dar al-Hilal, 1962).

Baqillani, Muhammad Ibn 'Abd al-Tayyib, *I'jaz al-Qur'an* (Cairo: Dar al-Ma'arif, 1954).

Bint al-Shati' (pseud.) ['A'ishah 'Abd al-Rahman], *Al-I'jaz al-Bayani li al-Qur'an* (Cairo: Dar al-Ma'arif, 1971).

Al-Dib, Muhammad, *Al-Bayan fi I'jaz al-Qur'an* (Cairo: Maktabat Muhammad 'Ali Sabih. 1960).

Al-Hufi, Ahmad Muhammad, *al-Mar'ah fi al-Shi'r al-Jahili*, 3rd edn. (Cairo: Dar al-Fikr al-'Arabi, 1963).

Husayn, 'Abd-al-Qadir, *Al-Qur'an: I'jazuhu wa Balaghatuhu* (Egypt: Maktabat al-Amanah, 1975).

Ibn Al-Jawzi, 'Abd al-Rahmin ibn 'Ali, *Ahkam al-Nisa'*, edited by 'Ali ibn Muhammad Yusuf al-Mahdi (Bayrut: Manshurat al-Maktabah al-Ajariyyah, 1981).

Khalaf-Allah, Muhammad Ahmad, *Al-Fann al-Qasasi fi al-Qur'an al-Karim* (Cairo, Egypt: Maktab al-Anjali Masriyyah, 1965).

Bint al-Shati' (pseud.) ['A'ishah 'Abd al-Rahman], *al-I'jaz al-Bayani li al-Qur'an* (Cairo: Dar al-Ma'arif, 1971).

Khattab, 'Abd-al-*Mu'izz, Ishruna Mar'ah fi al-Qur'an* (Cairo: Dar al-Hammami Li al-Taba'ah, 1970).

Al-Maraghi, Ahmad Mustafa, *'Ulum al-Balaghah: al-Bayan wa al-Ma'ani wa al-Badi'* (Egypt: al-Maktabah al-Mahmudiyyah al-Tijariyyah, 1963).

Matlub, Ahmad, *Al-Balaghah 'inda al-Sakkaki* (Baghdad: Maktabat al-Nahdah, 1964).

Nawfal, 'Abd-al-Razzaq, *Al-I'jaz al-Adadi li al-Qur'an al-Karim* (Cairo: Dar al-Sha'b, 1975).

Qutb, Sayyid, *Al-Adalah al-Ijtima'iyyah fi al-Islam* (Cairo: Dar al-Shuruq, 1975), English translation by John B. Hardie (New York: Octagon Books, 1980).

_____ *Al-Islam wa Mushkilat al-Hadarah* (Cairo: Dar al-Kitab al-'Arabi. Isa al-Babi al-Jalbani and Co., 1962).

_____ *Mashahid al-Qiyamah fi al-Qur'an* (Egypt: Dar al-Ma'arif, 1961).

Al-Rashid, 'Abd al-'Aziz, *Tahdhir al-Muslimun 'an Ittiba' Ghayr Sabil al-Mu'minin* (Baghdad: Matba'at al-Azhar, 1963).

Shaltut, Mahmud, *Al-Qur'an wa al-Mar'ah* (Cairo: Matba'at al-Nahdah, 1963).

Al-Sindi, al-Imam, *Sunan al-Nasa'i*, with an explanation by al-Hafiz al-Suyuti. (Bayrut: Dar al-Kitab al-'Arabi, n.d.).

Non-Arabic Sources

Ahmad, Leila. 'Feminism and the Feminist Movement in the Middle East, A Preliminary Exploration: Turkey, Egypt, Algeria, People's Democratic Republic of Yemen, *Women and Islam: Women's Studies International Forum Magazine*, Vol. 5 (1982): 153-68.

Holy Bible, King James Version (New York: American Bible Association, 1970).

Archer, John and Lloyd, Barbara, *Sex and Gender* (New York: Cambridge University Press, 1985).

Austin, John Longshaw, *How to do Things with Words*, edited by, J.O. Urmaon and Marina Sbisa, 2nd edn., (Oxford: Clarendon Press, 1975).

Baffoun, Alya. 'Women and Social Change in the Muslim Arab World', *Women and Islam: Women's Studies International Forum Magazine*, 5 (1982): 227-43.

Baveja, Malik Ram, *Woman in Islam*, translation of *Awrat awr Islami T'alim* (New York: Advent Books, 1981).

Becker, Alton. 'Biography of a Sentence: A Burmese Proverb.' 1976.

Bianchi, Eugene C. and Ruether, Rosemary Radford, *From Machisimo to Mutuality* (New York: Paulist Press, 1976).

Black, Max, *Models and Metaphors* (New York: Cornell Press, 1962).

Burke, Kenneth, *The Rhetoric of Religion* (Boston: Beacon Press, 1961).

—— *A Rhetoric of Motives* (Berkeley: University of California Press, 1969).

Carver, George A., *Aesthetics and the Problem of Meaning* (New Haven: Yale University Press, 1952).

Charlesworth, Maxwell J., *Problems of Religious Language* (New Jersey: Prentice-Hall, 1974).

Christian, David, 'A Masculine Mythology of Femininity' *Female Sexuality* (Ann Arbor, Michigan: The University of Michigan Press, 1970).

Cooper, Elizabeth, *The Harim and the Purdah: Studies of Oriental Women* (London: T. F. Unwin Ltd., 1915).

Devito, *Psychology of Speech and Language* (New York: Random House, 1970).

Dorph, Kenneth Jan, 'Islamic Law in Contempory North Africa: A Study of the Laws of Divorce in the Maghreb, *Women and Islam: Women's Studies International Forum Magazine*, 5 (1982): 169-82.

Esposito, John L., *Women in Islamic Family Law* (New York: Syracuse University Press, 1982).

Flew, Anthony, *God and Philosophy* (New York: Harcourt, Brace and World, 1966).

Glover, Elizabeth R. S., *Great Queens, Famous Rulers of the East* (London: Hutchinson and Co. Ltd., 1928).

Haddad, Yvonne Y., *Contemporary Islam and the Challenge of History* (New York: State University of New York Press, 1982).

_____ 'Eve: Islamic Image of Woman, *Women and Islam: Women's Studies International Forum Magazine*, 5 (1982): 135-44.

_____ 'Islam, Women and Revolution in Twentieth-Century Arab Thought' in *Women, Religion and Social Change* edited by Yvonne Y. Haddad and Ellison B. Findley (New York: State University of New York Press, 1985).

Haddad, Yvonne Y. and Findley, Ellison B., eds., *Women, Religion and Social Change* (New York: State University of New York Press, 1985).

Hassan, Riffat, 'Made from Adam's Rib: The Woman's Creation Question, *Al-Mushir Theological Journal of the Christian Study Centre, Rawalpindi, Pakistan* (Autumn 1985).

Al-Hibri, Aziza, 'A Story of Islamic Herstory: Or How did We ever get into this Mess?', *Women and Islam: Women's Studies International Forum Magazine*, 5 (1982): 193-206.

Holbrook, David, *Sex and Dehumanization* (New York: Pitman Publishing Company, 1972).

Houghton, Ross C., *Women of the Orient* (Cincinnati: Hitchcock and Walden; New York: Nelson and Phillips, 1977).

Illich, Ivan, *Gender* (New York: Pantheon Books, 1982).

Izutsu, Toshihiko, *God and Man in the Koran: Semantics of the Koranic Weltanschauung* (Tokyo: The Keio Institute of Culture and Linguistic Studies, 1964).

Jeffner, Andres, *The Study of Religious Language* (Bloomsbury, London: SCM Press Ltd., 1972).

Khalifa, Rashad, *Miracle of the Qur'an* (St. Louis, Missouri: Islamic Productions International, 1973).

Khadduri, Majid, *The Islamic Conception of Justice* (Baltimore, Maryland: Johns Hopkins University Press, 1984).

Kotb, Sayed, see Sayyid Qutb. *Social Justice in Islam*, translated by John B. Hardie (New York: Octagon Books, 1980).

Lakoff, Robin, *Language and Woman's Place* (New York: Octagon Books, 1976).

Levinson, Stephen C., *Pragmatics* (Cambridge, England: Cambridge University Press, 1983).

Lyalls, Charles, *Translations of Ancient Arabian Poetry* (New York: Columbia University Press, 1930).

Masson, D., *Le Coran et la Revelations Judeo-Chretienne*, 2 vols. (Paris: Maisonneuve, 1958).

Mernissi, Fatima, *Beyond the Veil: Male-Female Dynamics in a Modern Muslim Society* (New York: John Wiley & Sons., Halsted Press Division, 1975).

_____ 'Virginity and Patriarchy', *Women and Islam: Women's Studies International Forum Magazine*, 5 (1982): 183-91.

Minai, Naila, *Women in Islam: Tradition and Transition in the Middle East* (New York: Seaview Books, 1981).

Mir, Mustansir, *Thematic and Structural Coherence in the Qur'an: A Study of Islahi's Concept of Nazm* (Ann Arbor: University of Michigan Microfilms International, 1987).

al-Na'im, Abdullahi, *Towards an Islamic Transformation: Civil Liberties, Human Rights and International Law* (New York: Syracuse University Press, 1990).

Nasr, Seyyed Hossein, *Traditional Islam in the Modern World* (London: K. P. I. Limited, 1987).

Nicholson, Linda, *Gender and History: The Limits of Social Theory in the Age of the Family* (New York: Columbia University Press, 1986).

Olds, Sally Wendkos, *The Eternal Garden: Seasons of our Sexuality* (New York: Time Books, 1985).

Parkinson, Dilworth B., 'Agreement of Ba'd and Kull, *Al-'Arabiyya, Journal of the American Association of Teachers of Arabic* (Spring and Autumn 1975): 52-68.

Pilavoglu, Mehmed Kemal, *Tarihe ve Dinimize göre Kadin* (Ankara: Güven Matbaasi, 1964).

Procter-Smith, Majorie, *In Her Own Rite: Reconstructing Feminist Liturgical Tradition* (Nashville: Abingdon Press, 1990).

Rahbar, Daud, *God of Justice: A Study of the Ethical Doctrine of the Qur'an* (Leiden: E. J. Brill, 1960).

Rahman, Fazlur, *Major Themes in The Qur'an* (Chicago and Minneapolis: Bibliotheca Islamica, 1980).

_____ *Islam and Modernity, Transformation of an Intellectual Tradition* (Chicago: The University of Chicago Press, 1981).

De Riencourt, Amaury, *Sex and Power in History* (New York: D. McKay Co., 1974).

Roberts, Robert, *The Social Laws of the Qoran* (London: Williams and Norgate Ltd., 1925).

el-Saadawi, Nawal, 'Woman and Islam', *Women and Islam: Women's Studies International Forum Magazine*, 5 (1982): 193-206.

Sabbah, Fatna A., *Woman in the Muslim Unconscious*, translated by Mary Jo Lakeland from the French *La Femme dans L'inconscient Musulman* (New York: Pergamon Press, 1984).

Schmidt, Alvin J., *Veiled and Silenced: How Culture Shaped Sexist Theology* (Macon, Georgia: Mercer University Press, 1989).

Searle, John R., *Speech Acts, An Essay in the Philosophy of Language* (Cambridge, England: Cambridge University Press, 1969).

Shamma, S. H. A. L., *The Ethical System Underlying the Qur'an* (Tübingen: Hopfer Verlag, 1959).

Sharif, M. Raihan, *Islamic Social Framework* (Lahore: Orientalia, 1954).

Siddique, Kaukab, *The Struggle of Muslim Women* (Kingsville, Maryland: American Society for Education and Religion, Inc., 1986).

Siddiqi, Mohammed Mazharuddin, *Women in Islam* (Lahore: Institute of Islamic Culture, 1972).

Smith, Jane, 'Woman, Religion and Social Change in Early Islam', in *Women, Religion and Social Change*, edited by Yvonne Y. Haddad and Ellison B. Findley (New York: State University of New York Press, 1985).

Smith, Jane I. and Haddad, Yvonne Y., 'Eve: Islamic Image of Woman', *Women and Islam: Women's Studies International Forum Magazine*, 5 (1982): 135-45.

_____ 'The Islamic Understanding of Death and Ressurection, in *Women, Religion and Social Change* edited by Yvonne Y. Haddad and Ellison B. Findley, (New York: State University of New York Press, 1985).

_____ 'Women in the Afterlife: Islamic View as seen from the Qur'an and Traditions, *Journal of the American Academy of Religion*, 43 (1975): 39-50.

Smith, William Robertson, *Kinship and Marriage in Early Arabia*, edited by Stanley A. Cook (London: A. & C. Black, 1907).

115

Starrett, Barbara. 'The Metaphors of Power' in *The Politics of Women's Spirituality*, edited by Charlene Spetnak' (Garden City, New York: Doubleday Anchor Press, 1982).

Tarvis, Carol and Wade, Carole, *The Longest War: Sex Differences in Perspective*, 2nd edn. (Orlando: Harcourt Brace Jovanovich, 1984).

Tarvis, Carol, 'Stereotypes, Socialization and Sexism', *Beyond Sex Roles* (New York: West Publishing Company, 1977).

Topaloglu, Bekir, *Islamda Kadin* (Istanbul: Istanbül Imam Hatip Okulu Meshlek Desleri Ogretmeni, 1965).

Turnagil, A. Rasid, *Islam Cemiyetinin Esas Temelleri* (Istanbul: Abdullah Ishiklar, 1963).

Van Sommer, Annie, *Our Moslem Sisters*, 3rd edn. (New York: Young People's Missionary Movement, 1907).

Waddy, Charis, *Women in Muslim History* (New York: Longman, 1980).

Wan Daud, Wan Mohd Nor, *The Concept of Knowledge in Islam and Its Implications for Education in a Developing Country* (London: Mansell Publisher Limited, 1989).

Woodsmall, Ruth F., *Moslem Women Enter a New World* (New York: Roundtable Press Inc., 1936).

_____ *Study of the Role of Women in the Near East* (New York: International Federation of Business and Professional Women, 1956).

Zein-ed-Din, Nazirah, 'Removing the Veil and Veiling', translated from the 1928 edn. by Salah-Dine Hammoud for *Women and Islam: Women's Studies International Forum Magazine*, 5 (1982): 221-7.

Index

117